AUDIO VISUAL HAND BOOK

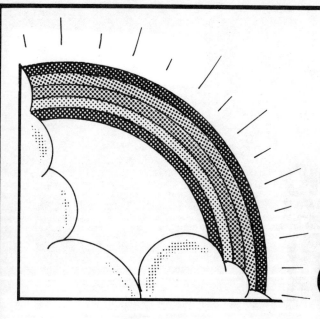

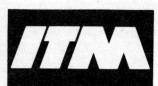

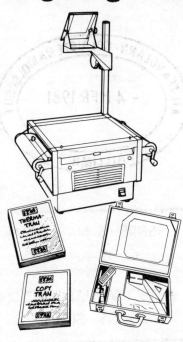

AUDIO VISUAL HANDBOOK

edited by
John Henderson and Fay Humphries

Kogan Page

Editors: John Henderson and Fay Humphries

Publisher's supervising editor: Loulou Brown

Research assistants: Carol Steiger, Leonie Farmer

Design: Jane Spicer

Production: Lesley Charlton, Kristine Hatch, Julia Talbut

First published 1980 as NAVAC Audio Visual Handbook
Published 1981 as The Audio Visual Handbook: The SCET Guide to Educational and Training Equipment

British Library Cataloguing in Publication Data
Henderson, John

The Audio Visual Handbook
I. Audio visual equipment
I. Title II. Humphries, Fay
621.38'044 TS230I.A7

ISBN 0-85038-454-0

Printed and bound in Great Britain by
The Anchor Press Ltd and bound by
Wm Brendon & Son Ltd, both of Tiptree, Essex

CONTENTS

Introduction 9

Products Directory

Equipment 12

Still image 12

Episcopes 12

Overhead projectors 16

Slide, filmstrip and slide/filmstrip viewers 26

Slide/filmstrip projectors 26

Slide projectors 30

Moving image 36

8mm cassette silent projectors 36

8mm cassette sound projectors 36

8mm spool silent projectors ... 37
8mm sound projectors ... 38
16mm film projectors ... 41

Audio ... 46
Amplified loudspeakers ... 46
Audio distribution units ... 47
Card readers ... 47
Cassette players ... 48
Cassette recorders ... 48
Language laboratories ... 53
Loudspeakers ... 55
Open reel tape recorders ... 57
Public address amplifiers ... 59
Radio cassette recorders ... 60
Radio receivers ... 62
Television receivers/monitors ... 63

Tape/visual ... 64
Dissolve units ... 64
Filmstrip/sound viewers ... 66
Slide/tape viewers ... 69
Synchronising cassette recorders ... 70

Video ... 72
Video cassette recorders ... 72
Video cameras ... 82

Screens ... 85
Front projection screens ... 85
Back projection screens ... 86

Non-projectional aids ... 87
Visual planning and display systems ... 87
Writing boards ... 87
Thermal copiers ... 88

Microcomputers ... 89

Manufacturers' names and addresses ... 92

Late entries ... 94

Media ... 95
Producers/distributors of filmstrips/slides with recordings ... 95
Producers/distributors of filmstrips/slide sets ... 96
Producers/distributors of overhead projector transparencies ... 98
Producers/distributors of 16mm films for education and training ... 98
Producers/distributors of videocassettes ... 100
Producers/distributors of audio cassettes ... 101
Producers/distributors of 8mm film loops ... 102
Producers/distributors of wallcharts ... 103
Producers/distributors of scientific models ... 104
Multi-media kits ... 104

Raw Software ... 105
Manufacturers/suppliers of photographic material ... 105
Manufacturers/suppliers of OHP materials ... 105
Manufacturers/suppliers of audio tapes ... 106
Manufacturers/suppliers of video tapes ... 106

Production services ... 108

Training and Information

Training 118
Higher degrees 118
Postgraduate diplomas 118
Other courses in educational technology and associated subjects 119

Professional organisations 121

List of film libraries 123

Professional and trade journals 124

Equipment hiring agencies 127

Selective bibliography 132
General audio visual education 132
Resources and resources centres 133
Media studies 134
Radio, television and video 135
Training and programmed learning 136
Photography 137
Reprography 137
Audio 138
Special subjects 138
Film 139

Index of advertisers 141

INTRODUCTION

This, the second edition of the *Audio Visual Handbook*, is designed, as was its predecessor, to help both the expert and the beginner to find their way through the morass of conflicting information and advice which exists with regard to audio visual equipment and services. It is primarily intended for the teacher, lecturer or trainer but will be equally useful to anyone involved in the communication of information or ideas in the business or industrial areas.

The first section is a directory of audio visual equipment and gives specifications of products which are currently available and generally suitable for educational purposes. In keeping with a policy of improvement this section has been changed in three major respects compared with the first edition.

(a) As far as possible all entries have been up-dated and at the same time many more pieces of equipment have been included. New categories have been added including 'screens', 'thermal copiers' and 'microprocessors'.

(b) The approximate price of each piece of equipment has been indicated using a banding system working in steps of £100 from £1 to £1000 and above. It should be remembered that all prices are subject to change, usually in an upward direction, and it follows that the price indicated for any product should only be regarded as a guide. The manufacturer or an accredited dealer should always be contacted to ascertain an exact cost before purchasing.

(c) A listing of manufacturers' names, addresses and telephone numbers has been compiled. This has been made as comprehensive as possible and the UK address has been given where possible.

Following the equipment directory are sections dealing with companies producing and supplying prepared materials for training and education, companies offering studio facilities and audio visual production facilities and manufacturers of raw software. Also included are details of both degree courses and short courses in many aspects of audio visual training and education. Lists of film libraries, equipment hire firms and professional organisations are also included.

While every care has been taken in the compiling of the various sections of the handbook it must be realised that the audio visual market is a rapidly expanding and changing one and the editors would like to emphasise the need for contacting manufacturers, firms, colleges and organisations listed to confirm any information. At the same time we would like to make it clear that any errors which have crept in are our responsibility. We apologise for these in advance and would be grateful if they were pointed out to us.

Our thanks to Mrs A Reid who meticulously and painstakingly checked every new or altered entry in the handbook.

PRODUCTS DIRECTORY

EQUIPMENT

The following section of the handbook gives specifications of currently available audio visual hardware. The items listed are those specifically designed for use in education and training with a proportion of those models which are designed for professional and domestic use where it was felt that they might be of interest because of price or the facilities provided. The following points regarding the entries should be noted:

1. It is assumed, unless stated otherwise, that all power supplies are AC.
2. The abbreviation 'nia' stands for 'no information available' and the editors regret the frequency with which it has had to be used.
3. A coding has been given in an attempt to indicate the cost of a particular model. The bands allocated are:

£0- £99 – (a)
£100-£199 – (b)
£200-£299 – (c)
£300-£399 – (d)
£400-£499 – (e)
£500-£599 – (f)
£600-£699 – (g)
£700-£799 – (h)
£800-£899 – (i)
£900-£999 – (j)
£1000 + – (k)

It must be realised that these price bands are based on the costs as available when the compiling was progressing and, considering present trends, it would be more than prudent to check with your supplier before making a firm decision about any particular model. The same is true with regard to any of the specifications. Manufacturers change and improve models continually, and editors make mistakes (occasionally). If any errors are spotted we would be grateful if they were pointed out to us.

STILL IMAGE

EPISCOPES

The episcope is designed to project images from opaque originals like photographs, the page of a book or, in some cases, three-dimensional objects. This is accomplished by directing light, from a high intensity source, on to the original. The light reflected from the original is then focused on the screen by a series of mirrors and lenses. As a result of this system the amount of light reaching the screen is considerably less than that from a comparable slide projector or similar device which uses transparent originals. The episcope is, therefore, only of use with an audience if the room being used has been completely blacked out, and even then some models are not very successful. However, the episcope is an ideal tool for producing enlarged sketches of a small original. If the small diagram is projected on to a sheet of card or paper the outline of the enlargement can be traced round, producing an accurate, but bigger, copy. The projection area varies considerably from model to model but is generally square. Depending on the design of the episcope some have a restricted thickness of original, while others will accept material of any size. The lamps used in episcopes vary considerably, from low voltage halogen lamps, through mains voltage to high power discharge lamps. They all, however, produce considerable amounts of heat when operating and the resulting high temperatures can damage delicate originals.

Make:	**Braun**
Model:	Paxiscope 650
Projection area:	145 x 145mm
Max copy thickness:	No limit – top loading
Lamp:	650W halogen
Lens:	200mm f/3.5
Optional lenses:	None
Power supply:	nia
Weight:	nia
Dimensions:	nia
Cost:	nia
Accessories:	None

Make:	**Demolux**
Model:	Demoscop

Projection area: 190 x 190mm
Max copy thickness: 60mm
Lamp: 1000W metal vapour
Lens: 260mm
Optional lenses: None
Power supply: 200/220/240W
Weight: 26kg
Dimensions: 580 x 340 x 490mm
Cost: nia
Accessories: None

Make: **Elite Optics**
Model: 91.3002 De Luxe
Projection area: 127 x 127mm
Max copy thickness: No limit – top loading
Lamp: Two 24V 150W tungsten halogen A1/216
Lens: 200mm, f/3.5
Optional lenses: None
Power supply: 240V
Weight: 7.8kg
Dimensions: 203 x 155 x 337mm
Cost: (a)
Accessories: None
Notes: Machine can be inverted to facilitate projection of large originals. Focusing by rotating projection lens

Make: **ITM**
Model: Primopaque
Projection area: 190 x 190mm
Max copy thickness: No limit – top loading (removable lid)
Lamp: 240V 1000W quartz-iodine P2/7
Lens: 400mm
Optional lenses: None
Power supply: 220-240V
Weight: 11.5kg
Dimensions: 368 x 304 x 444mm
Cost: (e)
Accessories: None
Notes: Focusing by single 'non-slip' lever

Make: **ITM**
Model: Vu-Lyte II
Projection area: 250 x 250mm
Max copy thickness: 50mm
Lamp: 650W quartz halogen
Lens: 432mm f/3.6
Optional lenses: None
Power supply: 220/240V
Weight: 15.5kg
Dimensions: 457 x 355 x 597mm
Cost: (h)
Accessories: None
Notes: Lens cover. Projection pointer highlights specific points on image. Focusing by large knurled knob. Vacuum platen – keeps material flat during projection

Make: **ITM**
Model: Vu-Lyte III
Projection area: 250 x 250mm
Max copy thickness: 50mm
Lamp: 650W quartz halogen
Lens: 432mm f/3.6
Optional lenses: None
Power supply: 220/240V
Weight: 16kg
Dimensions: 520 x 355 x 496mm
Cost: (h)
Accessories: None
Notes: Extra large access door allows projection of solid objects

Make: **Leitz**
Model: LE 19S
Projection area: 160 x 190mm
Max copy thickness: 55mm
Lamp: 220V 1000W metal halogen
Lens: 400mm EPIS f/4
Optional lenses: None
Power supply: 220V
Weight: 25kg
Dimensions: 790 x 265 x 435mm
Cost: (f)
Accessories: Slide for originals of A4 format; dust cover; carrying case
Other models: LE 19 – as above but 800W lamp, weight 19kg, dimensions 710 x 265 x 435mm, (d)
Notes: 1000W metal halogen lamp produces x4 brightness of 800W lamp, therefore projection can occur in only partial black-out

Make: **Liesegang**
Model: Antiskop
Projection area: 140 x 150mm
Max copy thickness: No limit
Lamp: 500W halogen floodlight – 2000 hrs
Lens: 200mm f/3.6
Optional lenses: 200mm f/3.5
Power supply: 220/240V
Weight: 6.2kg
Dimensions: 280 x 310 x 430mm
Cost: nia
Accessories: Copy holder for horizontal projection; vertical attachment with wall fastener; table clamp for vertical attachment; supplementary lens for reduction; water house stop and red filter for photographic work; dust cover; carrying case
Other models: Antiskop Super – as above with 400W halogen metal vapour lamp; 6000 hrs; permits working in normal lighting; lamp requires separate converter unit

Male: **Liesegang**
Model: Front Episcope
Projection area: 190 x 190mm
Max copy thickness: 55mm
Lamp: 1000W halogen metal vapour – 4000 hrs
Lens: Wide angle 3 element 260mm f/3.8
Optional lenses: None
Power supply: Separate power pack
Weight: 18kg. Power pack 9kg
Dimensions: 680 x 340 x 500mm. Power pack 150 x 250 x 220mm
Cost: (f)
Accessories: Dust cover, projection table
Notes: Focusing at handwheel via gearing system. Connection facility for Liesegang Fanti 150 slide projector. Retractable 190 x 300mm working area via slide base. If upper section lifted from slide base and placed directly on to specimen, deposit area is unlimited

Make: **Liesegang**
Model: E6
Projection area: 300 x 300mm
Max copy thickness: 55mm
Lamp: 2 x 1000W halogen floodlight
Lens: 480mm f/3.8
Optional lenses: 600mm f/3.8, 800mm f/4.5
Power supply: 220/240V
Weight: Depending on lens fitted – average 30kg

Dimensions:	Depending on lens fitted – average 740 x 435 x 555mm
Cost:	Depending on lens fitted: (j) or (k)
Accessories:	Light shielding curtain; dust cover; projection tables
Other models:	E6 Super Type 508 – as above with 2 x 1000W halogen metal vapour lamps and separate converter unit. This allows projection in only slightly dimmed rooms. Also available 1000m f/5.6 lens, cost (k) depending on lens.
Notes:	Special models eg for reproduction in natural size (1:1) or reduced scale, available on request

Make:	**Liesegang**
Model:	E8
Projection area:	190 x 160mm
Max copy thickness:	55mm
Lamp:	1000W halogen floodlight – 2000 hrs
Lens:	330mm f/3.6
Optional lenses:	400mm f/4.0
Power supply:	240V
Weight:	17kg
Dimensions:	580 x 380 x 410mm
Cost:	(d) to (e) with 400mm lens
Accessories:	Light shielding curtain; sliding copy holder; dust cover; carrying case; projection table
Other models:	E8 Super – as above but 400W halogen metal vapour – 6000 hrs – allows projection in semi-daylight; cost band (e) to (f) with 400mm lens

Make:	**Liesegang**
Model:	E9
Projection area:	190 x 160mm
Max copy thickness:	55mm
Lamp:	1000W halogen floodlight
Lens:	330mm f/3.6
Optional lenses:	400mm f/4.0
Power supply:	240V
Weight:	17kg
Dimensions:	700 x 340 x 410mm
Cost:	(e)
Accessories:	Dust cover; carrying case; projection table
Other models:	E9 Super as above with 400W halogen vapour lamp – 6000 hrs – allows projection in 'semi-daylight' conditions (f)
Notes:	E9 and E9 Super – the base tray slides underneath the unit to permit continuous projection of copies 190 x 310mm in size. For projection of even larger copies, eg maps, unit can be lifted off base tray and placed directly on to the material to be projected

Make:	**Liesegang**
Model:	E10 with separate converter unit
Projection area:	190 x 160mm
Max copy thickness:	55mm
Lamp:	1000W halogen metal vapour – 4000 hrs – allows 'semi-daylight' projection
Lens:	330mm f/3.6
Optional lenses:	400mm f/4.0 or 600mm f/3.8, or 800mm f/4.5 or 1000mm f/5.6
Power supply:	240V
Weight:	15.7kg choke unit 11.4kg
Dimensions:	560 x 410 x 330mm choke unit 275 x 240 x 150mm
Cost:	(f) (330 and 400mm lens)
Accessories:	Light shielding curtain; sliding copy holder; dust cover
Notes:	Copies 190 x 300cm may be projected continuously with accessory sliding copy holder

Make:	**Liesegang**
Model:	E11 Type 509 on sliding base tray
Projection area:	190 x 160mm
Max copy thickness:	55mm
Lamp:	1000W halogen metal vapour – 4000 hrs – semi-daylight projection
Lens:	330mm f/3.6
Optional lenses:	400mm f/4.0
Power supply:	240V
Weight:	17kg
Dimensions:	680 x 340 x 410mm
Cost:	(f)
Accessories:	Dust cover; projection table
Notes:	Sliding base tray as E9 permitting projection of larger copies

Make:	**Projection Optics**
Model:	Opa-scope TM
Projection area:	254 x 254mm
Max copy thickness:	63.5mm
Lamp:	1000W DRB
Lens:	457mm f/3.6 colour corrected
Optional lenses:	None
Power supply:	100-120V
Weight:	18.2kg
Dimensions:	508 x 355 x 584mm
Cost:	nia
Accessories:	Glass pressure plate; heat filter assembly (for use with heat sensitive materials); cover; platen plate; pad cover
Notes:	Built in 1000W optical pointer → brilliant 360° arrow; rack and pinion focusing; lens cover; vacuum platen – keeps material securely in place during projection

Make:	**THD**
Model:	Buhl Mark IV
Projection area:	254 x 254mm
Max copy thickness:	nia
Lamp:	240V 1000W
Lens:	457mm f/3.6
Optional lenses:	None
Power supply:	240V
Weight:	17.2kg
Dimensions:	330 x 520 x 533mm
Cost:	nia
Accessories:	None

OVERHEAD PROJECTORS

The overhead projector can be considered as the most useful piece of audio visual equipment available. As its name suggests, the OHP projects an image over the head (or shoulder) of the teacher. This means that the teacher need never turn his back on the class and is thus able to maintain eye contact at all times. This is useful for judging reactions to teaching points and keeping a good rapport with the audience.

The basic principle is that of projecting a bright image of whatever is placed on a glass stage. The projection lamp is housed below the stage and the light is focused by a system of lenses. The last part of the optical system is housed in a 'head' above the stage and the image is normally focused by raising or lowering this head. The head also has a mirror incorporated in it to enable the position of the image to be raised or lowered to a reasonable height for viewing. One disadvantage of the system is the formation of a distorted image if the top edge of the screen is not tilted forwards to compensate for 'keystoning'. Two stage sizes are in common use nowadays:

1. 250 x 250mm (10 inches square)
2. 285 x 285mm (A4 square)

The second size is the more recent and is especially useful if transparencies produced by copying techniques are to be used.

Generally, the lamp used will be of the low voltage halogen type, though some of the older designs use mains voltage lamps. Some projectors are equipped with a lamp changer mechanism which makes it possible to switch in a new lamp if the one in use should burn out.

The projector is normally used about 2 metres from the screen but many models have interchangeable heads which make it possible to bring the machine nearer to the screen for the same image size. This is especially useful in small classrooms.

Most models have what is called a 'scroll attachment'. This makes it possible to use a long roll of acetate film which can be wound across the stage as required.

Make:	**E J Arnold**
Model:	AV 290
Stage size:	255 x 255mm
Scroll:	N-S, E-W
Lamp:	24V 250W
Lamp changer:	No
Lens:	290mm
Optional lenses:	None
Cooling:	Fan
Power supply:	240V
Weight:	10.4kg
Dimensions:	240 x 340 x 400mm
Cost:	(b)
Accessories:	Scroll unit; trolley

Make:	**Bell & Howell**
Model:	1701
Stage size:	285 x 285mm
Scroll:	N-S, E-W
Lamp:	24V 250W tungsten halogen
Lamp changer:	Yes
Lens:	317mm
Optional lenses:	None
Cooling:	Fan
Power supply:	100-155V 200-265V
Weight:	nia
Dimensions:	nia
Cost:	1701 (b) 1702 (c) 1703 (c)
Accessories:	Scroll unit; accessory side tray; dust cover
Other models:	1702 (with condenser lens and built-in spare lamp). 1703 (with high definition triplet object lens, plano-convex condenser, anti-glare Fresnel and built-in spare lamp)
Notes:	Lamp life at full power: 40 hrs. 2 position switch doubles average life

Make:	**Bell & Howell**
Model:	Traveller 2304118
Stage size:	248 x 248mm with pen tray at side
Scroll:	No
Lamp:	24V 250W tungsten halogen A1/223
Lamp changer:	No
Lens:	Single element 290mm wide-angle
Optional lenses:	350mm 3 element; 315mm 3 element
Cooling:	Fan
Power supply:	240V (220V version also available)
Weight:	7.9kg (+4.4kg case)
Dimensions:	350 x 440 x 450mm; case 490 x 180 x 490mm
Cost:	(d)
Accessories:	Roll-feed attachment
Notes:	Lamp can be under run > 100 per cent in life. 24-hour digital clock. Auxiliary socket supply voltage outlet for ancillary equipment

Make:	**Clarke & Smith**
Model:	751E
Stage size:	260 x 260mm
Scroll:	N-S
Lamp:	230V 650W quartz iodide
Lamp changer:	Yes
Lens:	Dual lens facility 305mm and 355mm
Optional lenses:	None
Cooling:	Fan
Power supply:	240V (220V available on request)
Weight:	9.4kg
Dimensions:	360 x 375 x 570mm
Cost:	(b)
Accessories:	Dust cover
Notes:	Thermal cut-out. 2 position lamp switch for increased life

Make:	**Dixons**
Model:	AV 200 (portable with handle)
Stage size:	250 x 250mm
Scroll:	N-S, E-W
Lamp:	240V 65W tungsten halogen
Lamp changer:	No
Lens:	2 element 320mm
Optional lenses:	None
Cooling:	Fan
Power supply:	240V
Weight:	8.18kg
Dimensions:	368 x 317 x 260mm folded; height in use 679mm
Cost:	(b)
Accessories:	Scroll unit; dust covers

Make:	**Elf**
Model:	Scribe-Portable
Stage size:	255 x 255mm
Scroll:	N-S, E-W
Lamp:	24V 150W A1/216
Lamp changer:	No
Lens:	Single element 280mm
Optional lenses:	3 element
Cooling:	No fan

Power supply:	nia
Weight:	10kg
Dimensions:	255 x 310 x 727mm
Cost:	(b)
Accessories:	None
Notes:	2 position switch for increasing lamp life. Pen tray, scroll attachment and acetate roll are standard fitments

Make:	**Elf**
Model:	Superscribe
Stage size:	255 x 255mm
Scroll:	N-S, E-W
Lamp:	GE240DYR 650W QH
Lamp changer:	No
Lens:	2 element 355mm
Optional lenses:	305mm, 254mm
Cooling:	Fan
Power supply:	nia
Weight:	10kg
Dimensions:	Height 762mm
Cost:	(c)
Accessories:	None
Notes:	Two roll carriers, acetate roll and dust cover are included as standard fitments

Make:	**Elite Optics**
Model:	Viewrite 91.2017
Stage size:	285 x 285mm
Scroll:	N-S, E-W
Lamp:	24V 250W halogen A1/223
Lamp changer:	No
Lens:	2 element 355mm
Optional lenses:	310mm wide-angle (supplied in lieu at no extra cost)
Cooling:	Fan
Power supply:	230/250V (other voltages/frequencies available on request)
Weight:	11kg
Dimensions:	335 x 368 x 750mm
Cost:	(b)
Accessories:	Scroll unit; work support tables; anti-glare screen
Notes:	3 element high-definition lens will shortly be available as optional

Make:	**Elite Optics**
Model:	Viewrite 91.2028
Stage size:	254 x 254mm
Scroll:	N-S, E-W
Lamp:	24V 250W halogen A1/223
Lamp changer:	No
Lens:	2 element 355mm
Optional lenses:	310mm or 270mm
Cooling:	Fan
Power supply:	230-250V (other voltages/frequencies available on request)
Weight:	11kg
Dimensions:	335 x 368 x 750mm
Cost:	(b)
Accessories:	Scroll unit; work support tables; anti-glare screen
Other models:	91.2008 Portable
Notes:	Alternative lens supplied in lieu at no extra cost. 3 element lens as optional extra available when particularly fine detail is required

Make:	**Elite Optics**
Model:	Viewrite 91.2008 LVP
Stage size:	254 x 254mm
Scroll:	N-S, E-W
Lamp:	24V 250W halogen A1/223
Lamp changer:	No
Lens:	280mm
Optional lenses:	None
Cooling:	Fan
Power supply:	230-250V (other voltage/frequencies available on request)
Weight:	10.4kg
Dimensions:	335 x 368 x 670mm
Cost:	(b)
Accessories:	Carrying case; scroll unit; work support tables; anti-glare screen
Other models:	Viewrite 91.2020 LVP – as above with 285 x 285mm stage (b)
Notes:	Portability-angled steel column and lens head are secured with the instrument body for transportation. Carrying handle on front

Make:	**Elite Optics**
Model:	Elite 2000 91.2031
Stage size:	254 x 254mm
Scroll:	N-S, E-W
Lamp:	LV M331Q
Lamp changer:	No
Lens:	2 element 355mm
Optional lenses:	310mm or 270mm (can be supplied in lieu at no extra cost)
Cooling:	Fan
Power supply:	230/250V (other voltages/frequencies available on request)
Weight:	11kg
Dimensions:	335 x 368 x 750mm
Cost:	(b)
Accessories:	Scroll unit; work support tables; anti-glare screen
Other models:	Elite 2000 91.2032 – as above but stage size 285mm x 285mm (b)
Notes:	Economy 2 position lamp switch increases lamp life to ~ 300 hrs; 3 element lens available as optional extra when particularly fine detail is required. Optical tuning adjustment – operated by lever

Make:	**Elmo**
Model:	HP-A290
Stage size:	285 x 285mm
Scroll:	N-S, E-W
Lamp:	24V-250W EHJ
Lamp changer:	No
Optional lenses:	None
Cooling:	Fan
Power supply:	AC
Weight:	10.5kg
Dimensions:	355 x 470 x 230mm (folded) 780mm (assembled)
Cost:	(a)
Accessories:	Scroll unit
Notes:	Carrying handles enables complete portability

Make:	**Elmo**
Model:	HP 702
Stage size:	170 x 170mm
Scroll:	N-S
Lamp:	650W halogen
Lamp changer:	No
Lens:	220mm plus magnifying device (x 1.5)
Optional lenses:	Magnifying lens (x 1.5)
Cooling:	Fan
Power supply:	AC
Weight:	6.4kg
Dimensions:	255 x 270 x 500mm
Cost:	nia
Accessories:	Scroll unit

Notes: Designed for projecting small-size materials. Slides sizes 60 x 60mm, 70 or 90mm and 4'' x 5'' can be projected. The slide changer allows easy projection of successive slides

Make: **Elmo**
Model: HP 2450
Stage size: 254 x 254mm
Scroll: N-S, E-W
Lamp: 650W halogen
Lamp changer: Yes
Lens: 4 element 245mm wide-angle
Optional lenses: None
Cooling: Fan
Power supply: AC
Weight: 10.9kg
Dimensions: 314 x 352 x 766mm
Cost: (b)
Accessories: Scroll unit; pointer
Notes: Projects a large image at a very short distance. Built-in thermal cut-out to prevent overheating

Make: **Elmo**
Model: HP 2600
Stage size: 254 x 254mm
Scroll: N-S, E-W
Lamp: 650W halogen
Lamp changer: No
Lens: Wide-angle 260mm
Optional lenses: None
Cooling: Fan
Power supply: AC
Weight: 10.5kg
Dimensions: 314 x 352 x 656mm
Cost: (b)
Accessories: Scroll unit; pointer
Notes: Built-in thermal cut-out to prevent overheating

Make: **Elmo**
Model: HP 3000
Stage size: 254 x 254mm
Scroll: N-S, E-W
Lamp: 650W halogen
Lamp changer: Yes
Lens: 3 element 300mm
Optional lenses: None
Cooling: Fan
Power supply: AC
Weight: 11.2kg
Dimensions: 314 x 352 x 915mm
Cost: (b)
Accessories: Scroll unit; pointer
Notes: Built-in thermal cut-out

Make: **Elmo**
Model: HP 3300
Stage size: 254 x 254mm
Scroll: N-S, E-W
Lamp: 650W halogen
Lamp changer: No
Lens: 330mm
Optional lenses: None
Cooling: Fan
Power supply: AC
Weight: 9.8kg
Dimensions: 314 x 352 x 748mm
Cost: (b)
Accessories: Scroll unit; pointer
Notes: Built-in thermal cut-out

Make: **Erskine Westayr**
Model: 10 LV SE
Stage size: 254 x 254mm
Scroll: N-S, E-W
Lamp: 24V 250W tungsten halogen
Lamp changer: No
Lens: Single element 317mm
Options lenses: 274mm
Cooling: Fan
Power supply: 220/240V
Weight: 16.7kg
Dimensions: 381 x 394 x 692mm
Cost: (b)
Accessories: None
Other models: A4 LV SE – as above but 285 x 285mm stage (b)

Make: **Erskine Westayr**
Model: 10 LV
Stage size: 254 x 254mm
Scroll: N-S, E-W
Lamp: 24V 250W tungsten halogen
Lamp changer: No
Lens: 3 element 333mm
Optional lenses: None
Cooling: Fan
Power supply: 220/240V
Weight: 18.2kg
Dimensions: 381 x 394 x 692mm
Cost: (c)
Accessories: None
Other models: A4 LV 317mm lens, 385 x 385mm stage (c)

Make: **Erskine Westayr**
Model: Transepi 10 TLV
Stage size: 254 x 254mm
Scroll: N-S
Lamp: 24V 250W tungsten halogen
Lamp changer: No
Lens: 3 element 333mm OHP lens; 3 element 320mm EPI lens
Optional lenses: None
Cooling: Fan
Power supply: 220/240V or 110V
Weight: 24.05kg
Dimensions: 394 x 470 x 782mm
Cost: (e)
Accessories: Trolley
Other models: A4 TLV 284 x 284mm stage, 317mm lens
Notes: Sliding focus control for episcope lens. Dual purpose OHP and epidiascope

Make: **Fordigraph**
Model: 90
Stage size: 285 x 285mm
Scroll: N-S, E-W
Lamp: 24V 250W quartz halogen
Lamp changer: No
Lens: nia
Optional lenses: None
Cooling: Fan
Power supply: 220/240V or 110/115V
Weight: 11.34kg
Dimensions: 355 x 380 x 645mm
Cost: (b)
Accessories: Scroll unit
Other models: 80 – as above with 254 x 254mm stage (b). Model 70 as model 80 with lower light intensity lamp (b)
Notes: 2 position light switch for increased lamp life

Make: **Fordigraph**
Model: Inflight
Stage size: 285 x 285mm
Scroll: N-S
Lamp: 24V 250W quartz halogen
Lamp changer: Yes
Lens: 3 element
Optional lenses: None
Cooling: Fan
Power supply: 220-240V or 110-115V
Weight: 11kg
Dimensions: 365 x 365 x 115mm (collapsed)
Cost: (d)
Accessories: Scroll unit
Notes: Carrying case supplied with inner pocket for storing OHPTs

Make: **Ilado**
Model: 2004 NV
Stage size: 285 x 285mm
Scroll: N-S
Lamp: 36V 400W A1/239
Lamp changer: Yes
Lens: Single element 317mm
Optional lenses: None
Cooling: Fan
Power supply: nia
Weight: nia
Dimensions: 340 x 490 x 220mm (without objective lens column)
Cost: (d)
Accessories: Scroll unit; extension table attachment, glare shield, dust cover
Other models: 2004 D/NV – as above, but with dimmer control (see note below) (d). 2004 3/NV-SK – as above with 3 element 315mm lens (d). 2004 3D/NV-SK – as 2004 3/NV-SK but with dimmer control (d)
Notes: 2 position economy lamp switch for increased lamp life on models 2004 NV and 2004 NV-SK. Infinitely variable dimmer control switch on models 2004 D/NV and 2004 3D/NV-SK. Dimmer control can be fitted as optional extra to the other two models on request

Make: **ITM**
Model: Portascribe 330 LV
Stage size: 254 x 254mm
Scroll: N-S, E-W
Lamp: 24V 250W tungsten halogen A1/223. Low voltage
Lamp changer: No
Lens: Single element 320mm
Optional lenses: None
Cooling: Fan
Power supply: 240V
Weight: 11.9kg
Dimensions: 335 x 365 x 787mm
Cost: (b)
Accessories: Scroll unit; dust cover
Other models: Portascribe 430 LV – stage size A4 (285 x 285mm)
Notes: Low voltage lamp has higher tolerance to rough treatment

Make: **ITM**
Model: Portascribe 335/4
Stage size: 254 x 254mm
Scroll: N-S, E-W

Lamp: 24V 250W tungsten halogen A1/223. Low voltage
Lamp changer: No
Lens: 2 element 355mm
Optional lenses: 2 element 317mm; 2 element 266mm
Cooling: Fan
Power supply: 240V
Weight: 12.2kg
Dimensions: 355 x 365 x 787mm
Cost: (c)
Accessories: Scroll unit; dust cover
Other models: Portascribe 435 – stage size A4 (285 x 285mm)
Notes: Portascribe 435 – available with 317mm lens or 266mm lens only (c). Equipped with double-lamp change facility

Make: **ITM**
Model: Portascribe 700
Stage size: 254 x 254mm
Scroll: N-S, E-W
Lamp: 650W 240V tungsten halogen
Lamp changer: No
Lens: 2 element 355mm
Optional lenses: 2 element 317mm; 2 element 266mm
Cooling: Fan
Power supply: 240V
Weight: 9.8kg
Dimensions: 266 x 330 x 787mm
Cost: (c)
Accessories: Scroll unit; dust covers
Notes: Lamp economy switch which reduces light intensity by 10 per cent but increases the lamp life by up to 100 per cent

Make: **ITM**
Model: Portascribe 2000 A4
Stage size: 295 x 295mm
Scroll: N-S, E-W
Lamp: 125V 2000W quartz ref FEY rated life 500 hours
Lamp changer: No
Lens: 330mm
Optional lenses: Special lens available on request
Cooling: 2 turbine fans
Power supply: 240V
Weight: 26kg plus table 7kg
Dimensions: 400 x 540 x 1100mm (with table)
Cost: (i)
Accessories: Scroll unit; anti-glare shield
Notes: A variac control for regulating the voltage to the lamp from 0 to max is incorporated. Interlock switch prevents the lamp being switched on when in the maximum light position. A lamp elevation system operates from outside the projector. Table supplied, detractable wheels ensuring stability when in use. Can project X-ray plates

Make: **ITM**
Model: Portascribe LVA4
Stage size: 285 x 285mm
Scroll: N-S, E-W
Lamp: 24V 250W tungsten halogen A1/223. Low voltage
Lamp changer: Yes
Lens: 2 element 300mm condenser lens
Optional lenses: None
Cooling: Fan
Power supply: 240V
Weight: 15.5kg
Dimensions: 394 x 394 x 787mm
Cost: nia
Accessories: Scroll unit; dust cover
Notes: External optical tuning to eliminate colour distortion. Lamp economy dimmer switch

Make: **ITM**
Model: Portable 2401 LV with case
Stage size: 254 x 254mm
Scroll: No
Lamp: 24V 250W A1/223 tungsten halogen. Low voltage
Lamp changer: No
Lens: Single element, 300mm
Optional lenses: 3 element 315mm
Cooling: Fan
Power supply: 240V
Weight: 8kg
Dimensions: 350 x 450 x 450mm
Cost: (d) (single element lens)
Accessories: None
Other models: Portable 2403 LV with case – 3 element 315mm lens (d)
Notes: Dimmer switch incorporated giving up to 200 per cent increase in lamp life. Built-in digital clock which can be set to time of day or to zero position at beginning of lecture

Make: **ITM**
Model: Master Radiograph
Stage size: 406 x 330mm
Scroll: No
Lamp: 115V 1500W
Lamp changer: No
Lens: 940mm
Optional lenses: None
Cooling: Fan
Power supply: 200/240V
Weight: 68kg
Dimensions: nia
Cost: nia
Accessories: None
Notes: Variac operated voltage control. Masking shutter. High magnification lens

Make: **Kindermann**
Model: Famulus 2 6110
Stage size: 285 x 285mm
Scroll: N-S, E-W
Lamp: 24V 250W halogen
Lamp changer: No
Lens: 2 element 355mm
Optional lenses: 1 element 300mm; 3 element 300mm
Cooling: nia
Power supply: 220V
Weight: nia
Dimensions: nia
Cost: (c)
Accessories: Scroll unit; side table; polarizing spinner
Other models: 6115 – as above but with rapid lamp changer. 6130 as 6110 with 3 element 300mm lens head. 6135 as 6130 with rapid lamp changer. 6140 as 6130 with single element 300mm lens head. 6145 as 6140 with rapid lamp changer
Notes: Lamp house adjustment with knob outside machine. Economy position lamp switch gives 1000 hrs lamp life. Scroll unit and lamp not supplied as standard

Make: **Leitz**
Model: Diascriptor OP 2500
Stage size: 285 x 285mm
Scroll: N-S

Lamp:	24V 250W tungsten halogen
Lamp changer:	No
Lens:	285mm
Optional lenses:	340mm
Cooling:	Fan
Power supply:	220V
Weight:	14.3kg
Dimensions:	nia
Cost:	(b)
Accessories:	Scroll unit; anti-dazzle plate; dust cover; accessory case; attachment shelf
Other models:	OP 2500 with 340mm lens (c)
Notes:	Economy 2 position lamp switch, trebles life (> 150 hours)

Make:	**Liesegang**
Model:	601
Stage size:	285 x 285mm
Scroll:	N-S, E-W
Lamp:	900W halogen
Lamp changer:	No
Lens:	340mm
Optional lenses:	315mm, 290mm
Cooling:	Fan
Power supply:	220V
Weight:	10kg
Dimensions:	370 x 370 x 700mm
Cost:	nia
Accessories:	Glare shield; scroll unit; lateral shelf; dust cover
Other models:	603 – as above but 250 x 250mm stage plus projection head (355mm) available. 610 as 601 with 36V 400W LV halogen lamp. 611 as 610 with 250mm x 250mm stage
Notes:	Dimmer switch for prolonged life. 36V/400W LV lamp can be fitted. Illumination adjustable from outside

Make:	**Liesegang**
Model:	608
Stage size:	285 x 285mm
Scroll:	N-S, E-W
Lamp:	24V 250W
Lamp changer:	No
Lens:	315mm
Optional lenses:	290mm, 340mm
Cooling:	Fan
Power supply:	220V (110V version available)
Weight:	12.5kg
Dimensions:	370 x 370 x 700mm
Cost:	(c)
Accessories:	Scroll unit; glare shield; lateral shelf; dust cover
Other models:	609 – as above – stage 250 x 250mm
Notes:	Dimmer switch increases lamp life by X4. Illumination adjustable from outside. Telescoping column optional

Make:	**Liesegang**
Model:	614 Portable plus case
Stage size:	250 x 250mm
Scroll:	No
Lamp:	24V 250W
Lamp changer:	No
Lens:	Single element 300mm
Optional lenses:	3 element 315mm
Cooling:	Fan
Power supply:	220V (110/220V model available

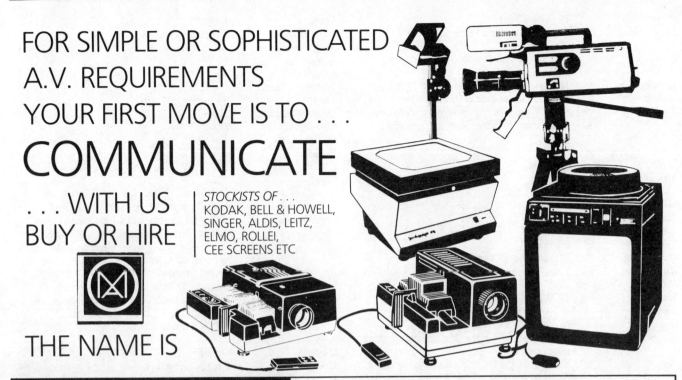

Weight: 8kg (case 3kg)
Dimensions: 350 x 440 x 450mm (case 500 x 500 x 200mm)
Cost: (d) (Single element lens) (d) (3 element 315mm lens)
Accessories: Dust cover; glare shield; cleaning set
Notes: Lamp dimmer switch increases lamp life by x 4. Digital clock. Safety plug socket for connection of additional units eg slide projector

Make: **Liesegang**
Model: 618 QC
Stage size: 285 x 285mm
Scroll: N-S, E-W
Lamp: 24V 250W halogen A1/223
Lamp changer: Yes
Lens: 290mm
Optional lenses: 315mm and 340mm
Cooling: Fan
Power supply: 220V
Weight: 12.5kg
Dimensions: 370 x 370 x 700mm
Cost: (b)
Accessories: Scroll unit; anti-glare shield; lateral tables; dust covers
Notes: The top of the focusing column can have fitted to it the Fanti 150 35mm projector which allows the incorporation of 35mm colour slides, film strips or microscope slides into the presentation

Make: **Lumatic**
Model: HA 224
Stage size: 254 x 254mm
Scroll: N-S
Lamp: 24V 250W A1/233 halogen
Lamp changer: No
Lens: Standard 290mm
Optional lenses: 2 element 355mm; 3 element 317mm
Cooling: Turbo fan
Power supply: nia
Weight: 13kg
Dimensions: 310 x 340 x 280mm (without column)
Cost: (b)
Accessories: Scroll unit; magnifier; dust covers
Other models: HA 224-A4 – as above but stage size 295 x 295mm
Notes: HA 224 S modified to USPEC 3 requirements

Make: **Ormig**
Model: Grapholux 202/1
Stage size: 250 x 250mm
Scroll: N-S
Lamp: 24V 320W
Lamp changer: No
Lens: Single element 310mm
Optional lenses: None
Cooling: Fan
Power supply: nia
Weight: nia
Dimensions: 369 x 369 x 268mm (excluding column)
Cost: (c)
Accessories: Carrying handles standard. Scroll unit; side tray; anti-glare shield; polarizing filter
Other models: 202/2 – as above but with 2 element 355mm lens slide attachment for 50 x 50mm glass slides as optional extra. 202/3 as 202/1 but with 3 element 317mm lens. Models 202/1, 2, 3 are also available in A4 format, ie as described above but with 300 x 300mm stage

Make: **Ormig**
Model: Grapholux 401/3
Stage size: 250 x 250mm
Scroll: N-S
Lamp: 36V 480W
Lamp changer: No
Lens: 3 element 317mm
Optional lenses: None
Cooling: Fan
Power supply: nia
Weight: nia
Dimensions: 369 x 369 x 268mm (without column)
Cost: nia
Accessories: Carrying handle; side trays; scroll unit; anti-glare shield; polarizing filter
Other models: 402/1 – as above but single element, 310mm lens, stage size 300 x 300mm 402/2 – as 402/1 but 2 element 355mm lens slide attachment (for 50 x 50mm glass slides) available. 402/3 – as 402/1 but 3 element 317mm lens
Notes: Additional economy lamp switch on all above models increases lamp life to 400-500 hours

Make: **Ormig**
Model: Grapholux 501/1
Stage size: 250 x 250mm
Scroll: N-S
Lamp: 240V 460W
Lamp changer: No
Lens: Single element 310mm
Optional lenses: None
Cooling: Fan
Power supply: nia
Weight: nia
Dimensions: 369 x 369 x 268mm (excluding column)
Cost: (c)
Accessories: Carrying handles; side trays; scroll units; anti-glare shield
Other models: 502/1 – as above but 300 x 300mm stage (c)
Notes: Projection lamp case knob control on front of machine

Make: **Ormig**
Model: Grapholux 801/2
Stage size: 250 x 250mm
Scroll: N-S
Lamp: 220V 820W
Lamp changer: No
Lens: 2 element 355mm
Optional lenses: None
Cooling: Fan
Power supply: nia
Weight: nia
Dimensions: 369 x 369 x 268mm (excluding column)
Cost: (c)
Accessories: Carrying handles standard; side trays; scroll unit; slide attachment; anti-glare shield; polarizing filter
Notes: Prolonged lamp life with additional economy circuit

Make: **Ormig**
Model: Grapholux 1002/3
Stage size: 300 x 300mm
Scroll: N-S
Lamp: 220V 1030W halogen
Lamp changer: No
Lens: 3 element, 317mm wide-angle
Optional lenses: None
Cooling: Fan
Power supply: nia

Weight: nia
Dimensions: 369 x 369 x 268mm (excluding column)
Cost: nia
Accessories: Carrying handles standard; side trays; scroll unit; anti-glare shield; polarizing filter
Notes: Prolonged lamp life from additional lamp economy circuit

Make: **Paul Plus**
Model: Plus A4
Stage size: 285 x 285mm
Scroll: N-S, E-W
Lamp: 24V 250W tungsten halogen
Lamp changer: No
Lens: 300mm f3.3
Optional lenses: None
Cooling: Turbo type fan
Power supply: Max 286W
Weight: 14kg
Dimensions: 345 x 380 x 285mm (excluding focusing column)
Cost: (b)
Accessories: Scroll unit
Other models: Plus 25 – as above, 250 x 250mm stage size (b)

Make: **Rank Aldis**
Model: Rank Aldis OHP
Stage size: 285 x 285mm or 254 x 254mm
Scroll: nia
Lamp: 24V 250W halogen
Lamp changer: No
Lens: Single element 275mm
Optional lenses: 3 element 315mm, extra condenser
Cooling: Fan
Power supply: 220/240V
Weight: 14kg
Dimensions: 345 x 370 x 900mm
Cost: (b)
Accessories: Dust cover; scroll unit

Make: **3M**
Model: 213S
Stage size: 254 x 254mm
Scroll: N-S, E-W
Lamp: 82V 360W 5 per cent, ASA Code ENX
Lamp changer: No
Lens: 355mm
Optional lenses: None
Cooling: Fan
Power supply: 240V
Weight: 9.7kg
Dimensions: 356 x 399 x 660mm
Cost: (b)
Accessories: Scroll unit; side tables
Notes: External colour tuning

Make: **3M**
Model: 213
Stage size: 254 x 254mm
Scroll: No
Lamp: 82V 360W plus 5 per cent ASA Code ENX
Lamp changer: Yes
Lens: 355mm
Optional lenses: None
Cooling: Fan
Power supply: 240V
Weight: 9.7kg
Dimensions: 356 x 399 x 660mm
Cost: (c)
Accessories: Side tables
Notes: 2 position lamp switch – increase of 2½ on life. Optical tuning adjustment

Make: **3M**
Model: 213 LC
Stage size: 266 x 266mm
Scroll: N-S, E-W
Lamp: 82V 360W ASA Code ENX
Lamp changer: Yes
Lens: 355mm
Optional lenses: None
Cooling: Fan
Power supply: 240V
Weight: 12.7kg
Dimensions: 381 x 406 x 680mm
Cost: (c)
Accessories: Side tables; scroll unit
Notes: Safety thermostat cuts power to lamp if cooling system blocked. External colour tuning – eliminates red/blue corners. 2 position lamp switch for increased life. Pen tray

Make: **3M**
Model: 213 Portable
Stage size: 267 x 266mm
Scroll: N-S, E-W
Lamp: 82V 360W 5 per cent ASA Code ENX
Lamp changer: Yes
Lens: 355mm
Optional lenses: None
Cooling: Fan
Power supply: 240V
Weight: 12.7kg
Dimensions: 381 x 306 x 680mm
Cost: (c)
Accessories: Scroll attachment
Notes: Optical tuning adjustment. 2 position lamp switch increases lamp life by factor of 2½

Make: **3M**
Model: 213 Triplet
Stage size: 286 x 286mm
Scroll: No
Lamp: 82V 360W plus or minus 5 per cent ASA Code ENX
Lamp changer: Yes
Lens: Triplet lens, 317mm
Optional lenses: None
Cooling: Fan
Power supply: 240V
Weight: 12.7kg
Dimensions: 381 x 406 x 680mm
Cost: (d)
Accessories: Side tables
Notes: Optical tuning adjustment. 2 position lamp switch > x 2½ increase in life. Pen tray

Make: **3M**
Model: 0-88 Portable with carrying case
Stage size: 266 x 266mm
Scroll: No
Lamp: Single filament 500W 240V (500W 120V also available) 75 hour life rating
Lamp changer: No
Lens: Single moniscus configuration
Optional lenses: None
Cooling: No fan: convection louvres
Power supply: 240V
Weight: 4.30kg (plus case – 4.77kg)
Dimensions: 324 x 349 x 457mm
Cost: (c)
Accessories: None

Make: **Weyel**
Model: 79810
Stage size: 285 x 285mm

Scroll: N-S, E-W
Lamp: 24V 250W halogen
Lamp changer: Yes
Lens: 355mm
Optional lenses: None
Cooling: nia
Power supply: nia
Weight: nia
Dimensions: 390 x 510 x 880mm
Cost: (d)
Accessories: Scroll unit (260mm or 300mm)
Notes: Projection head can also be turned horizontally. 2 position lamp switch extends lamp life. A special long-life lamp (1000hrs on economy circuit, 300hrs on maximum output) is available as an optional extra

SLIDE/FILMSTRIP AND SLIDE/FILMSTRIP VIEWERS

The units listed below can be used to view slide or filmstrip or both. The general design projects the image onto an integral back projection screen and, since the screens are not large, the viewing group cannot be greater than two or three and they are really intended for individual use.

Make: **Gateway**
Model: Prima Slide Viewer
Slide size: 50 x 50mm slides
Slide transport: Manual, 24 slide stack loader
Screen size: 127 x 127mm
Lamp: 14V, 25W
Power supply: 240V
Weight: 1.75kg
Dimensions: 143 x 283 x 289mm
Cost: (a)
Accessories: Two-slide shuttle changer

Make: **Gateway**
Model: Superviewer
Filmstrip size: 35mm, single frame
Filmstrip transport: Spool
Screen size: 127 x 171mm
Lamp: 110V, 50W
Power supply: 240V
Weight: 4kg
Dimensions: 280 x 280 x 305mm
Cost: (a)
Accessories: None

Make: **Gateway**
Model: Prima Filmstrip Viewer
Filmstrip size: 35mm, single frame
Filmstrip transport: Lever operated pawl
Screen size: 89 x 127mm
Lamp: 14V, 25W
Power supply: 240V
Weight: 1.3kg
Dimensions: 143 x 273 x 283mm
Cost: (a)
Accessories: None

Make: **Gateway**
Model: Cabin 900A
Filmstrip size: 35mm, double frame
Filmstrip transport: Spools
Slide size: 50 x 50mm
Slide transport: Manual
Magazine type: None
Screen size: 104 x 104mm
Lamp: 230V, 150W
Lens: nia
Power supply: 240V
Weight: 2.2kg
Dimensions: 120 x 260 x 320mm
Cost: (a)
Accessories: Multi-slide auto-changer; carrying case

SLIDE/FILMSTRIP PROJECTORS

These projectors are designed to enable the user to project either slides, of varying sizes and formats, or filmstrips, both half-frame and full-frame. This is accomplished by having interchangeable carriers which can be attached between the lamp and the lens system of the projector. Generally the loading is manual but a few machines have a stack loading device for slides which makes it possible to use them as semi-automatic devices.

The lamp type varies from machine to machine, some having low voltage halogen lamps with a high light output while others, more suited to small group use, have low output lamps.

Make: **Bell & Howell**
Model: Dinoscop
Filmstrip size: Single and double frame
Filmstrip transport: Manual
Slide size: 50 x 50mm
Slide transport: Manual
Magazine type: None
Lamp: 24V, 250W (A1/223)
Lens: 100mm f/2.8
Optional lenses: 375mm; 150mm; 85-150mm zoom
Power supply: 220-240V
Weight: 6.5kg
Dimensions: 460 x 165 x 115mm
Cost: (a)
Accessories: Carrying case

Make: **Dukane**
Model: 28A 57A
Filmstrip size: Single frame
Filmstrip transport: Remote
Slide sizes: 50 x 50mm
Slide transport: Manual
Magazine type: None
Lamp: 500W
Lens: 125mm f/2.8
Optional lenses: 38mm f/2.8; 76mm f/2.5
Power supply: 120V
Weight: 5kg
Dimensions: 254 x 240 x 133mm
Cost: (b)
Accessories: Slide changer; carrying case; remote extension cable; cover
Other models: Micromatic II, as above with sync control socket (c)

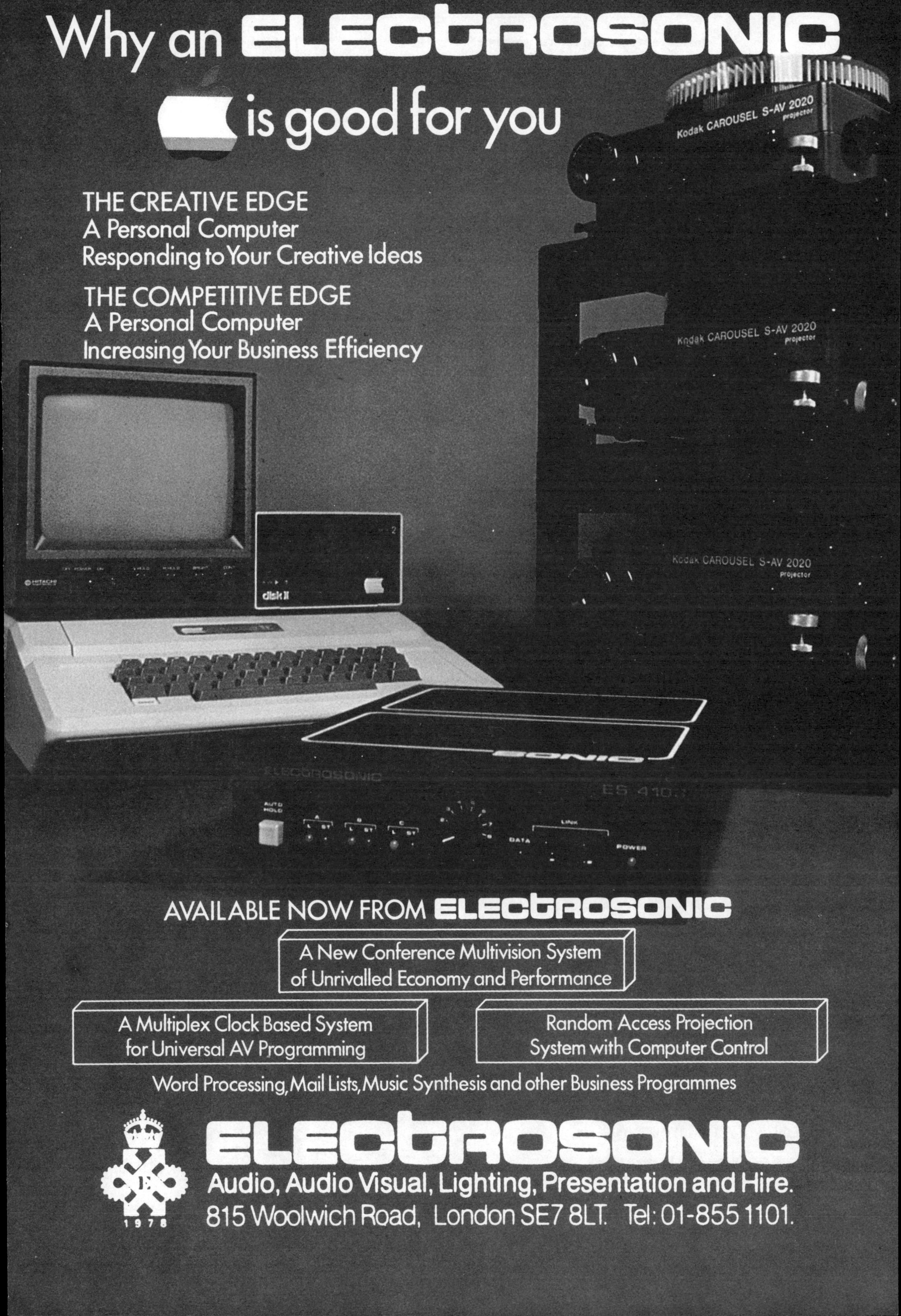
Why an ELECTROSONIC
is good for you
THE CREATIVE EDGE
A Personal Computer
Responding to Your Creative Ideas
THE COMPETITIVE EDGE
A Personal Computer
Increasing Your Business Efficiency
Kodak CAROUSEL S-AV 2020 projector
disk II
AVAILABLE NOW FROM ELECTROSONIC
A New Conference Multivision System of Unrivalled Economy and Performance
A Multiplex Clock Based System for Universal AV Programming
Random Access Projection System with Computer Control
Word Processing, Mail Lists, Music Synthesis and other Business Programmes
ELECTROSONIC
Audio, Audio Visual, Lighting, Presentation and Hire.
815 Woolwich Road, London SE7 8LT. Tel: 01-855 1101.
1978

Notes: Filmstrip advance and reverse is remotely controlled at two speeds, one frame per activation of switch or 200 frames in 45s

Make: **Elmo**
Model: S-300
Filmstrip size: Single and double frame
Filmstrip transport: Manual
Slide size: 50 x 50mm
Slide transport: Semi-automatic
Magazine type: None
Lamp: 240V, 300W
Lens: 85mm f/2.8
Optional lenses: 50mm f/2.8
Power supply: 110-125V; 220-240V
Weight: 1.5kg
Dimensions: 300 x 175 x 95mm
Cost: (a)
Accessories: Carrying case; conversion lens giving 68mm or 106mm; film strip attachment; semi-auto changer
Notes: Preview screen to enable slide sorting

Make: **Hanimex**
Model: Hanimette Carrel 2
Filmstrip size: Single and double frame
Filmstrip transport: Manual
Slide size: 50 x 50mm
Slide transport: Manual
Magazine type: None
Lamp: 240V, 100W (A1/21)
Lens: 58mm f/3.5
Optional lenses: None
Power supply: 100-250V (according to lamp) AC/DC
Weight: 1.25kg
Dimensions: 255 x 130 x 100mm
Cost: nia
Accessories: Rotary slide carrier

Make: **Hanimex**
Model: Syllabus 4000
Filmstrip size: Single or double frame
Filmstrip transport: Manual
Slide size: 50 x 50mm
Slide transport: Manual
Magazine type: None
Lamp: 24V, 250W (A1/223)
Lens: 100mm f/2.8
Optional lenses: 150mm; 75-125mm zoom
Power supply: 230/240V
Weight: 6.8kg
Dimensions: 360 x 245 x 145mm
Cost: nia
Accessories: Semi-automatic magazine slide changer; micro slide attachment; rotary slide carrier
Notes: Two light level switches

Make: **Leitz**
Model: Prado Universal 5 x 5
Filmstrip size: Single and double frame
Filmstrip transport: Manual
Slide size: 50 x 50mm
Slide transport: Manual
Magazine type: None
Lamp: 24V, 250W (A1/223)
Lens: 90mm f/2.5
Optional lenses: 35mm to 300mm
Power supply: 110-125V adjustable
Weight: 7.7kg
Dimensions: 320 x 195 x 150mm
Cost: nia
Accessories: Carrying case; filmstrip attachment

Make: **Liesegang**
Model: Fantimat 150
Filmstrip size: Single and double frame
Filmstrip transport: Manual
Slide size: 50 x 50mm
Slide transport: Remote
Magazine type: Leitz straight (36 or 50)
Lamp: 24V, 150W
Lens: 85mm f/2.8
Optional lenses: 50mm f/2.8; 100mm f/2.8; 150mm f/3.5; 200mm f/3.5; 250mm f/4.2; 70-120mm f/3.5 zoom
Power supply: 110/130/220/240V
Weight: 4.1kg
Dimensions: 306 x 260 x 120mm
Cost: (a)
Accessories: Filmstrip attachment; micro projector; infrared remote control
Other models: Fantimat 150AF, as above with autofocus, timer and dissolve socket (b)

Make: **Liesegang**
Model: Fantimat 250
Filmstrip size: Single and double frame
Filmstrip transport: Manual
Slide size: 50 x 50mm
Slide transport: Remote
Magazine type: Leitz straight (36 or 50)
Lamp: 24V, 250W
Lens: 85mm f/2.8
Optional lenses: See Fantimat 150
Power supply: 110/130/220/240V
Weight: 4.3kg
Dimensions: 306 x 260 x 120mm
Cost: (b)
Accessories: See Fantimat 150
Other models: Fantimat 250AF, as above with autofocus, timer and dissolve socket (b)

Make: **Malinverno**
Model: SAV 1000
Filmstrip size: Single and double frame
Filmstrip transport: Manual
Slide size: 50 x 50mm
Slide transport: Manual
Magazine type: None
Lamp: 240V, 100W (A1/21)
Lens: 85mm f/2.8
Optional lenses: 40mm; 50mm; 100mm; 85-150mm; 200mm
Power supply: 240V
Weight: 2kg
Dimensions: 305 x 170 x 115mm
Cost: nia
Accessories: Semi-automatic magazine slide changer; micro slide attachment; filmstrip carrier; carrying case
Other models: SAV 2000, as above with 12V, 50W lamp. SAV 4000, as above with 25V, 150W lamp and optional motorised filmstrip carrier

Make: **Malinverno**
Model: SAV 5000
Filmstrip size: Single and double frame
Filmstrip transport: Manual
Slide size: 50 x 50mm
Slide transport: Manual
Magazine type: None
Lamp: 24V, 250W (A1/223)
Lens: 100mm f/2.8
Optional lenses: 40mm; 50mm; 85mm; 135mm; 200mm; 85-150mm zoom
Power supply: 110/125/220/240V

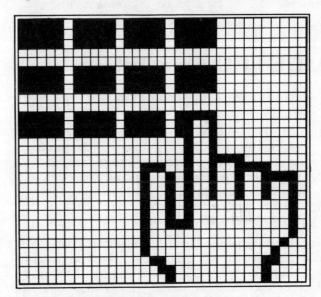

Weight: 7.8kg
Dimensions: 385 x 222 x 200mm
Cost: nia
Accessories: See SAV 4000

Make: **Prestinox**
Model: Prestige 2000
Filmstrip size: Single and double frame
Filmstrip transport: Manual
Slide size: 50 x 50mm
Slide transport: Manual
Magazine type: None
Lamp: 24V, 150W
Lens: 100mm f/2.8
Optional lenses: 60mm; 85mm; 150mm; 200mm; 70-120mm zoom
Power supply: 115-230V
Weight: nia
Dimensions: nia
Cost: nia
Accessories: Semi-automatic slide changer
Other models: Elysee 2000, as above with facility for 60 x 60mm and 50 x 50mm slides

Make: **Rank Aldis**
Model: Tutor 2
Filmstrip size: Single and double frame
Filmstrip transport: Manual
Slide size: 50 x 50mm
Slide transport: Manual
Magazine type: None
Lamp: 24V, 250W
Lens: 100mm f/2.8
Optional lenses: 60mm; 85mm; 150mm
Power supply: 110/130/150/220/240/250V
Weight: 8.6kg
Dimensions: 500 x 225 x 112mm
Cost: (b)
Accessories: Semi-automatic slide changer; carrying case

Notes: External 12V socket for operating accessories

Make: **THD**
Model: Halight 24/250
Filmstrip size: Single and double frame
Filmstrip transport: Manual
Slide size: 50 x 50mm
Slide transport: Manual
Magazine type: None
Lamp: 24V, 250W (A1/223)
Lens: 85mm f/2.8
Optional lenses: 60mm; 100mm; 70-120mm zoom
Power supply: 240V
Weight: 6.4kg
Dimensions: 350 x 190 x 110mm
Cost: (b)
Accessories: See Halight 300

Make: **THD**
Model: Halight 300
Filmstrip size: Single and double frame
Filmstrip transport: Manual
Slide size: 50 x 50mm
Slide transport: Manual
Magazine type: None
Lamp: 240V, 300W (A1/249)
Lens: 85mm f/2.8
Optional lenses: 60mm; 100mm; 70-120mm zoom
Power supply: 220-240V
Weight: 2.8kg
Dimensions: 330 x 130 x 100mm
Cost: (a)
Accessories: Microscope/microfiche reader attachment; polarising filters; rotating stage; water cell; graticule; carrying case
Other models: Halight 100B, as above with 12V, 100W lamp for battery operation (a)

SLIDE PROJECTORS

There are three main sizes of slide produced by modern cameras:

1. 110 format cameras produce slides which are normally mounted in a 30 x 30mm mount but these can be remounted in special 50 x 50mm mounts.
2. 35mm format cameras give slides which are 24 x 36mm and are mounted in 50 x 50mm mounts.
3. Roll film cameras usually require the slides to be mounted in 60 x 60mm mounts.

Most of the projectors in this section require the slides to be loaded into some form of magazine or tray. There are two types, straight and rotary, and there are various designs of each type. These are not normally interchangeable. The three most common patterns are:

1. The Leitz universal pattern, straight (36 or 50 slides) and rotary (100 slides). This pattern is used by a number of European manufacturers.
2. The Hanimex/Gnome pattern, straight (36 slides) and rotary (100 or 120 slides). This style is used by various machines and is common to both Gnome and Hanimex projectors.
3. The Kodak Carousel pattern, rotary only (80 slides).

Slide changes can be carried out in various ways:

(a) *Manual.* The projector will have no magazine and each slide is loaded individually by the operator.
(b) *Semi-automatic.* The slides are in a magazine. Each slide is placed in the projection position by means of a sliding mechanism moved by the operator.
(c) *Automatic.* The slides are in a magazine and the slide loading is carried out by a mechanism controlled by push buttons on the projector.
(d) *Remote.* This type of slide change is possible with automatic projectors if they are equipped with a socket into which a lead and control unit can be plugged.
(e) *Infra-red remote control.* This uses special control units which do away with the need for the connecting cable.
(f) *Tape synchronising.* On many automatic projectors the remote control socket allows the use of a synchronising lead which connects to a tape or cassette recorder and enables automatic control for tape/slide sequences.
(g) *Fade and dissolve.* Some projectors have special sockets and associated circuitry to fade and dissolve units or multivision systems to be connected.

Most projectors use low voltage halogen lamps which give a

greater light output than a similar wattage mains voltage lamp.

The various lenses available have two parameters:

1. *The focal length (in mm)*. The greater this is the further the projector can be from the screen for the same image size. Zoom lenses have a continuously variable focal length between two limits and are useful where the projector is going to be used in a variety of situations with different projector to screen distances.
2. *The aperture (f number)*. The smaller this number is the greater the amount of light which is allowed through the lens.

Automatic focusing, which is now available on some projectors, ensures that all the slides in a sequence stay in focus throughout, once the initial focusing has been carried out.

Make:	**Agfa-Gevaert**
Model:	Pocket projector
Slide size:	30 x 30mm
Slide transport:	Automatic
Magazine type:	Agfa-Leitz rotary
Sockets:	None
Lamp:	12V, 75W
Lens:	45mm f/2.8
Optional lenses:	None
Power supply:	110, 116, 130, 220, 240V
Weight:	3.9kg
Dimensions:	292 x 230 x 134mm
Cost:	(a)
Accessories:	None
Notes:	Remote control; light beam pointer; single slide projection facility

Make:	**Agfa-Gevaert**
Model:	Diamotor 1500
Slide size:	50 x 50mm
Slide transport:	Automatic
Magazine type:	Agfa CS (40 or 100); Agfa-Leitz (36 or 50)
Sockets:	Sync
Lamp:	24V, 150W
Lens:	85mm f/2.8
Optional lenses:	60mm f/2.8; 90mm f/2.5; 150mm f/3.5
Focus:	Remote
Power supply:	nia
Weight:	nia
Dimensions:	nia
Cost:	(a)
Accessories:	None
Other models:	Diamotor 1500 AF, as above with autofocus (a)

Make:	**Bergen**
Model:	Quadrapoint LV
Slide size:	50 x 50mm
Slide transport:	Automatic
Magazine type:	Carousel rotary (80)
Sockets:	Remote/sync; dissolve
Lamp:	Xenon arc lamp
Lens:	nia
Optional lenses:	nia
Power supply:	220-240V
Weight:	16kg
Dimensions:	520 x 292 x 203mm
Cost:	nia
Accessories:	None

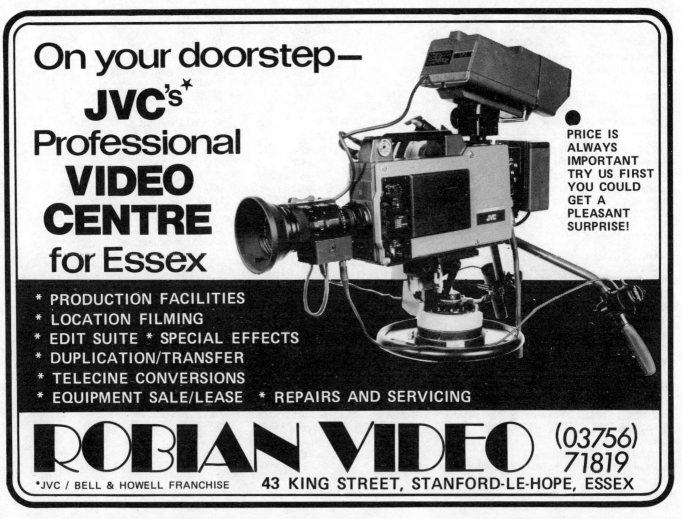

Make: **Braun**
Model: Novamat 520
Slide size: 50 x 50mm
Slide transport: Automatic
Magazine type: Leitz
Sockets: Remote
Lamp: 24V, 150W
Lens: 85mm f/2.5
Optional lenses: None
Power supply: nia
Weight: nia
Dimensions: nia
Cost: nia
Accessories: Dissolve unit for Novamat 540 AE only
Other models: Novamat 540A, as above with automatic control. Novamat 540 AE, as 540A with light intensity control

Make: **Braun**
Model: Paximat 750
Slide size: 50 x 50mm
Slide transport: Automatic
Magazine type: Leitz straight (36 or 50), rotary (100)
Sockets: Remote/sync
Lamp: 24V, 150W
Lens: 85mm f/2.8
Optional lenses: 55mm f/2.5; 90mm f/2.5; 100mm f/2.8; 150mm f/3.5; 70-120mm f/3.5 zoom
Focus: Remote/manual
Power supply: nia
Weight: 4.5kg
Dimensions: 257 x 262 x 136mm
Cost: nia
Accessories: None
Other models: Paximat 950, as above with autofocus

Make: **Braun**
Model: Paximat 5150
Slide size: 50 x 50mm
Slide transport: Automatic
Magazine type: Leitz straight (36 or 50), rotary (100)
Sockets: Remote/sync
Lamp: 24V, 150W
Lens: 85mm f/2.8
Optional lenses: See Paximat 750
Focus: Remote
Power supply: nia
Weight: 5.15kg
Dimensions: 280 x 295 x 135mm
Cost: nia
Accessories: None
Other models: Paximat 5150A, as above with autofocus. Paximat 5150AE, as with 5150A variable light control

Make: **Braun**
Model: Paximat 5250AE
Slide size: 50 x 50mm
Slide transport: Automatic
Magazine type: Leitz straight (36 or 50), rotary (100)
Sockets: Remote/sync; dissolve
Lamp: 24V, 250W
Lens: 85mm f/2.8
Optional lenses: See Paximat 750
Focus: Automatic
Power supply: nia
Weight: 6.6kg
Dimensions: 280 x 295 x 135mm
Cost: nia
Accessories: Dissolve unit
Other models: Paximat 5250 AEF, as above with infra-red remote control

Make: **Enna-Werk**
Model: Ennamat
Slide size: 50 x 50mm
Slide transport: Automatic
Magazine type: Leitz straight (36 or 50)
Sockets: Sync
Lamp: 24V, 150W (A1/216)
Lens: See below
Optional lenses: 45mm to 150mm
Focus: Remote
Power supply: nia
Weight: 4kg
Dimensions: 280 x 245 x 120mm
Cost: (a)
Accessories: None
Other models: Various models available incorporating economiser circuit; timer; autofocus, in various combinations

Make: **GAF**
Model: 101
Slide size: 50 x 50mm
Slide transport: Semi-automatic
Magazine type: Leitz
Sockets: None
Lamp: 24V, 150W
Lens: 85mm f/2.8
Optional lenses: None
Focus: Manual
Power supply: nia
Weight: 3.8kg
Dimensions: 265 x 240 x 110mm
Cost: (a)
Accessories: None

Make: **GAF**
Model: 501
Slide size: 50 x 50mm
Slide transport: Automatic/semi-automatic
Magazine type: Leitz straight (36 or 50), rotary (100)
Sockets: Remote/sync
Lamp: 24V, 150W
Lens: 85mm f/2.8
Optional lenses: 60mm f/2.8; 150mm f/3; 70-120mm f/3.5 zoom
Focus: Remote/manual
Power supply: nia
Weight: nia
Dimensions: nia
Cost: (a)
Accessories: Remote control
Other models: 502, as above with autofocus and economy lamp switch (a). 503, as 502 with timer and facility for synchronising with stereo tape recorder

Make: **Gnome**
Model: 722 Alpha major
Slide size: 60 x 60mm
Slide transport: Manual
Magazine type: None
Sockets: None
Lamp: 24V, 150W (A1/216)
Lens: 150mm f/3.5
Optional lenses: None
Focus: Manual
Power supply: nia
Weight: 5.4kg
Dimensions: 318 x 165 x 95mm
Cost: (a)
Accessories: Adaptor for 50 x 50mm slides; lenses for 50 x 50mm slides

Other models: 723 Viscount, as above with 240V 150W (A1/243) lamp (a)

Make: **Gnome**
Model: 786 Sprite
Slide size: 50 x 50mm
Slide transport: Manual
Magazine type: None
Sockets: None
Lamp: 240V, 150W (A1/182)
Lens: 85mm f/2.8
Optional lenses: None
Focus: Manual
Power supply: nia
Weight: 1.4kg
Dimensions: 260 x 128 x 102mm
Cost: (a)
Accessories: None

Make: **Gnome**
Model: 7031 Sprite magazine
Slide size: 50 x 50mm
Slide transport: Semi-automatic
Magazine type: Hanimex/Gnome straight (36), rotary (120)
Sockets: None
Lamp: 12V, 50W (A1/220)
Lens: 85mm f/2.8
Optional lenses: None
Focus: Manual
Power supply: nia
Weight: 1.92kg
Dimensions: 273 x 198 x 114mm
Cost: (a)
Accessories: None
Other models: 7032 Sprite magazine IQ – as above with 240V, 150W lamp (a)

Make: **Gnome**
Model: 7303 Jaguar II
Slide size: 50 x 50mm
Slide transport: Semi-automatic
Magazine type: Hanimex/Gnome straight (36), rotary (120)
Sockets: None
Lamp: 24V, 150W (A1/216)
Lens: 85mm f/2.8
Optional lenses: 200m and interchangeable lenses available
Focus: Manual
Power supply: nia
Weight: 5kg
Dimensions: 290 x 275 x 130mm
Cost: (a)
Accessories: None
Other models: 7304 Jaguar II, as above with remote slide change and sync socket (a). 7306 Jaguar II, as 7304 with remote focus (a). 7309 Jaguar II, as 7306 with economy lamp switch (a)

Make: **Hanimex**
Model: Rondette 1500 RF
Slide size: 50 x 50mm
Slide transport: Automatic
Magazine type: Hanimex/Gnome straight (36 or 42), rotary (120)
Sockets: Sync/timer
Lamp: 24V, 150W (A1/216)
Lens: 85mm f/2.8
Optional lenses: 70-120mm zoom
Focus: Remote
Power supply: 240V
Weight: 4.4kg
Dimensions: 300 x 265 x 120mm
Cost: nia
Accessories: 3-30 S timer
Other models: Rondette 1500 EF, as above with autofocus

Make: **Imatronic**
Model: Planet CF 66 AF-S
Slide size: 60 x 60mm
Slide transport: Automatic
Magazine type: Straight
Sockets: Remote/sync
Lamp: 24V, 250W
Lens: 135mm f/2.8
Optional lenses: 150mm; 200mm; 250mm
Focus: Automatic
Power supply: nia
Weight: nia
Dimensions: nia
Cost: (c)
Accessories: Remote control; dissolve unit; carrying case
Notes: Economy lamp switch

Make: **Kindermann**
Model: 1800
Slide size: 50 x 50mm
Slide transport: Automatic
Magazine type: Leitz straight (36 or 50), rotary (100)
Sockets: None
Lamp: 150W
Lens: 90mm f/2.5
Optional lenses: 50mm; 85mm; 100mm; 150mm; 70-120mm zoom; 85-150mm zoom
Focus: nia
Power supply: nia
Weight: nia
Dimensions: nia
Cost: (a)
Accessories: None
Other models: 1480, as 1800 with 250W lamp (b). 1810, as 1800 with autofocus and timer (b). 1490, as 1810 with infra-red remote control (b)
Notes: Economy switch

Make: **Kindermann**
Model: 1870
Slide size: 50 x 50mm
Slide transport: Automatic
Magazine type: Leitz straight (36 or 50), rotary (100)
Sockets: Dissolve
Lamp: 150W
Lens: 90mm f/2.5
Optional lenses: See 1800
Focus: nia
Power supply: nia
Weight: nia
Dimensions: nia
Cost: (b)
Accessories: Dissolve control
Other models: 1890 – as above with 250W lamp (b)
Notes: Economy switch

Make: **Kindermann**
Model: 1899 Traveller
Slide size: 50 x 50mm
Slide transport: Manual
Magazine type: Leitz straight (36 or 50)
Sockets: None
Lamp: 12V, 20W
Lens: 85mm f/2.8
Optional lenses: None
Focus: Manual
Power supply: 12V DC (car battery)
Weight: 2.5kg

Dimensions: 276 x 255 x 125mm
Cost: (a)
Accessories: None

Make: **Kindermann**
Model: Telefocus 66
Slide size: 60 x 60mm
Slide transport: Automatic
Magazine type: Straight
Sockets: Remote/sync
Lamp: 250W
Lens: nia
Optional lenses: nia
Focus: Remote
Power supply: nia
Weight: nia
Dimensions: nia
Cost: (b)
Accessories: Dissolve unit; infra-red remote control
Other models: Super 66 – as above without remote control facility
Notes: Economy switch

Make: **Kodak**
Model: Carousel S-AV 1000
Slide size: 50 x 50mm
Slide transport: Automatic
Magazine type: Carousel rotary (80)
Sockets: Remote/sync
Lamp: 24V, 150W (A1/216)
Lens: See below
Optional lenses: 35mm to 250mm; 70-120mm zoom
Focus: Remote/manual
Power supply: 110-250V
Weight: 6.2kg
Dimensions: 280 x 280 x 100mm
Cost: (b)
Accessories: Remote control; remote control extensions; timer; carrying case
Other models: S-AV 2020 – as above with 24V, 250W (A1/223 lamp) 12pm socket for dissolve unit; spare lamp switch over (c)

Make: **Kodak**
Model: Carousel S-RA 2000
Slide size: 50 x 50mm
Slide transport: Automatic
Magazine type: Carousel rotary (80)
Sockets: Remote/sync
Lamp: 24V, 250W (A1/223)
Lens: See below
Optional lenses: 35mm to 250mm; 70-120mm zoom
Focus: Remote/manual
Power supply: 110-240V
Weight: 6.2kg
Dimensions: 280 x 280 x 100mm
Cost: (e)
Accessories: Remote control; keyboard control; carrying case; timer
Notes: Random access projector, any slide can be selected within 1.5-5s

Make: **Leitz**
Model: Pradovit R150
Slide size: 50 x 50mm
Slide transport: Automatic
Magazine type: Leitz
Sockets: Remote/sync
Lamp: 24V, 150W
Lens: 90mm f/2.5
Optional lenses: 50mm f/2.8; 85mm f/2.8; 120mm f/2.8; 150mm f/3.2
Focus: Remote
Power supply: 220V or 110-240V
Weight: 4.3kg
Dimensions: 279 x 262 x 132mm
Cost: (b)
Accessories: Carrying case
Other models: Pradovit RA 150 – as above with autofocus (b)

Make: **Leitz**
Model: Pradovit C 1500
Slide size: 50 x 50mm
Slide transport: Automatic
Magazine type: Leitz straight (36 or 50)
Sockets: Remote/sync
Lamp: 24V, 150W (A1/216)
Lens: 85mm f/2.8
Optional lenses: 35mm f/2.8; 50mm f/2.8; 90mm f/2.5; 120mm f/2.5; 150mm f/2.8; 200mm f/3.6; 250mm f/4
Focus: Remote
Power supply: 110-240V
Weight: 8kg
Dimensions: 334 x 268 x 166mm
Cost: (c)
Accessories: Infra-red remote control; timer; carrying case
Other models: Pradovit CA 1500 – as above with autofocus (c)
Notes: Economy lamp switch, remote light pointer; special slide-change mechanism reduces slide-change interval to 0.4s

Make: **Leitz**
Model: Pradovit C 2500
Slide size: 50 x 50mm
Slide transport: Automatic
Magazine type: Leitz (36 or 50)
Sockets: Remote/sync
Lamp: 24V, 250W
Lens: 85mm f/2.8
Optional lenses: See Pradovit C 1500
Focus: Remote
Power supply: 110-240V
Weight: 8kg
Dimensions: 334 x 268 x 166mm
Cost: (c)
Accessories: See Pradovit C 1500
Other models: Pradovit CA 2500 – as above with autofocus (c)
Notes: See Pradovit C 1500

Make: **Liesegang**
Model: Fantax 600 M
Slide size: 60 x 60mm
Slide transport: Semi-automatic
Magazine type: Leitz
Sockets: None
Lamp: 24V, 250W
Lens: 150mm f/3
Optional lenses: 200mm f/3.5; 250mm f/4.2; 150mm f/2.8
Focus: Manual
Power supply: 110, 130, 220, 240V
Weight: 5.3kg
Dimensions: 332 x 262 x 184mm
Cost: (b)
Accessories: None
Notes: Economy lamp switch

Make: **Liesegang**
Model: Fantax 600 A
Slide size: 60 x 60mm
Slide transport: Automatic
Magazine type: Leitz
Sockets: Remote/sync

Lamp: 24V, 250W
Lens: 150mm f/3
Optional lenses: 200mm f/3.5; 250mm f/4.2; 150mm f/2.8
Focus: Remote
Power supply: 110, 130, 220, 240V
Weight: 5.8kg
Dimensions: 332 x 262 x 184mm
Cost: (c)
Accessories: Remote control; infra-red remote control; dissolve unit for 600 A (d)
Other models: Fantax 600 A (d), as above with dissolve facility (c)
Notes: Economy lamp switch

Make: **Prestinox**
Model: 680 FE
Slide size: 50 x 50mm
Slide transport: Automatic
Magazine type: Leitz
Sockets: Remote/sync; dissolve
Lamp: 24V, 150W
Lens: 85mm f/2.8
Optional lenses: See 680 Auto
Focus: Automatic
Power supply: nia
Weight: nia
Dimensions: nia
Cost: nia
Accessories: Carrying case; dissolve unit
Notes: Safety cut-off switch; pointer light beam

Make: **Prestinox**
Model: 680 Auto
Slide size: 50 x 50mm
Slide transport: Automatic
Magazine type: Leitz
Sockets: Remote/sync
Lamp: 24V, 150W
Lens: 85mm f/2.8
Optional lenses: 50mm f/2.8; 100mm f/2.8; 125mm f/3.5; 150mm f/3.5; 200mm f/3.5; 70-120mm f/3.5 zoom
Focus: Remote
Power supply: nia
Weight: nia
Dimensions: nia
Cost: nia
Accessories: Carrying case
Other models: 680 autofocus – as above with autofocus and timer
Notes: Safety cut-off switch; pointer light beam

Make: **Prestinox**
Model: Dia-Technic AVP
Slide size: 50 x 50mm
Slide transport: Automatic
Magazine type: Leitz
Sockets: Remote/sync; dissolve
Lamp: 24V, 150W
Lens: 85mm f/2.8
Optional lenses: See 680 Auto
Focus: Automatic
Power supply: 110-220V
Weight: 5.8kg
Dimensions: 300 x 230 x 180mm
Cost: nia
Accessories: Remote control; dissolve unit
Notes: Timer; safety cut-off switch

Make: **Philips**
Model: Dia 20E
Slide size: 50 x 50mm
Slide transport: Automatic
Magazine type: Leitz
Sockets: Remote
Lamp: 24V, 150W (A1/216)
Lens: 85mm f/2.8
Optional lenses: 55mm f/2.8; 150mm f/3.5; 70-120mm zoom
Focus: Remote
Power supply: 110, 127, 220, 240V
Weight: nia
Dimensions: nia
Cost: (a)
Accessories: None
Other models: Dia 40A – as above with autofocus (b)
Notes: Economy lamp switch

Make: **Rollei**
Model: P355A
Slide size: 50 x 50mm
Slide transport: Automatic
Magazine type: Leitz
Sockets: Remote
Lamp: 24V, 150W (A1/216)
Lens: 85mm f/2.8
Optional lenses: 50mm f/2.8; 150mm f/3.5; 250mm f/4.3; 70-120mm f/3.5 zoom
Focus: Remote
Power supply: 110-240V
Weight: 4.5kg
Dimensions: 270 x 227 x 120mm
Cost: (a)
Accessories: Carrying case
Other models: P355 AF – as above with autofocus (a)

Make: **Rollei**
Model: P360A
Slide size: 50 x 50mm
Slide transport: Automatic
Magazine type: Leitz
Sockets: Remote; dissolve
Lamp: 24V, 250W
Lens: 85mm f/2.8
Optional lenses: See P355A
Focus: Remote
Power supply: 110-240V
Weight: 6.9kg
Dimensions: 328 x 293 x 123mm
Cost: (b)
Accessories: Carrying case; infra-red remote control
Other models: P360 AF, as above with autofocus (c)

Make: **Rollei**
Model: P3800 AF
Slide size: 50 x 50mm
Slide transport: Automatic
Magazine type: Leitz
Sockets: Sync
Lamp: 2 x 24V, 150W
Lens: 2 x 85mm f/2.8
Optional lenses: None
Focus: Automatic
Power supply: 110-240V
Weight: nia
Dimensions: nia
Cost: (d)
Accessories: Carrying case
Notes: A two-track two-lens projector, with built-in cross-fade unit

Make: **Rollei**
Model: P 665
Slide size: 60 x 60mm
Slide transport: Automatic
Magazine type: Straight
Sockets: Remote/sync
Lamp: 24V, 150W

Lens: 105mm f/3.5
Optional lenses: 180mm; 250mm; 400mm; 110-160mm zoom
Focus: Automatic
Power supply: nia
Weight: nia
Dimensions: nia
Cost: nia
Accessories: None

Make: **Simda**
Model: 2200
Slide size: 50 x 50mm
Slide transport: Automatic
Magazine type: Carousel rotary (80)
Sockets: Remote/sync; dissolve
Lamp: 24V, 250W
Lens: 85mm
Optional lenses: 100mm; 70-120mm zoom
Focus: Remote
Power supply: 200-250V
Weight: 7.4kg
Dimensions: nia
Cost: (d)
Accessories: Dissolve unit; remote control; carrying case
Other models: Simda 2400 – as above with 36V, 400W lamp (i)
Notes: Professional A-V presentation unit

MOVING IMAGE

8mm CASSETTE SILENT PROJECTORS

These projectors operate with the film contained in a plastic cassette. As a result, the film is never touched but the maximum running time is restricted to about four minutes. The film is loaded as an endless loop which makes the system ideal for repetitive viewing and the teaching of a single concept. Since the introduction of Super 8 film some projectors have been made which can cope with this format and the older Standard 8 machines are now being phased out.

Make: **Technicolor**
Model: 280
Film format: Super 8
Lamp: 30V, 80W
Lens: 20-30mm f/1.4 zoom
Optional lenses: None
Speed: 18 fps
Still picture: Yes
Power supply: 220-240V
Weight: 3kg
Dimensions: 240 x 180 x 140mm
Cost: nia
Accessories: None

Make: **THD**
Model: 205R
Film format: Super 8
Lamp: 24V 150W
Lens: 20-32mm f/1.4 zoom
Optional lenses: None
Speed: 18 fps
Still picture: Yes
Power supply: 220 or 240V
Weight: 6.4kg
Dimensions: 230 x 220 x 130mm
Cost: nia
Accessories: Rear projection cabinet
Other models: 105R – as above but standard 8 format

8mm CASSETTE SOUND PROJECTORS

Only Super 8 film is used in this type of cassette but the sound track can be either magnetic or optical. The cassettes are much larger than those used in silent machines and can accommodate up to 20 minutes of film.

Since the sound track is recorded on the film, much more complex ideas can be conveyed than with the silent cassette. Each manufacturer has tended to produce his own style of cassette and as a result they are not interchangeable.

Make: **Fairchild**
Model: Galaxy 900
Film format: Fairchild sound loop cartridge
Max playing time: 23 min
Screen size: 203 x 267mm
Lamp: 19V, 80W
Lens: nia
Optional lenses: None
Speed: 24 fps
Still picture: No
Sound head: Magnetic
Amplifier: 24W
Loudspeaker: 51 x 153mm, eliptical
Power supply: nia
Weight: 7kg
Dimensions: 419 x 318 x 406mm
Cost: nia
Accessories: Folding unit
Other models: Galaxy 900FA – as above with still picture and remote control

Make: **Fairchild**
Model: 70-07
Film format: Fairchild sound loop cartridge
Max playing time: 23 min
Screen size: 290 x 227mm, or front projection

Lamp:	19V, 80W
Lens:	10mm f/1.1
Optional lenses:	None
Speed:	24 fps
Still picture:	No
Sound head:	Magnetic
Amplifier:	3W
Loudspeaker:	50 x 250mm, eliptical
Power supply:	240V
Weight:	7.5kg
Dimensions:	340 x 445 x 140mm
Cost:	nia
Accessories:	Folding unit

Make:	**Technicolor**
Model:	Showcase
Film format:	Technicolor sound loop cartridge
Max playing time:	29 min
Screen size:	260 x 190mm or front projection
Lamp:	21V, 80W
Lens:	nia
Optional lenses:	None
Speed:	24 fps
Still picture:	Yes
Sound head:	Magnetic or optical available
Amplifier:	2W
Loudspeaker:	55 x 133mm, eliptical
Power supply:	220-240V
Weight:	10kg
Dimensions:	380 x 130 x 370mm
Cost:	nia
Accessories:	None
Other models:	2050E, 2050ME, 2000, 2000M; 2050M Shoemate, as Showcase with 280 x 210mm screen, 2100ME, 2100M

Make:	**Technicolor**
Model:	1100
Film format:	Technicolor sound loop cartridge
Max playing time:	29 min
Screen size:	nia, front projection possible
Lamp:	21V, 150W
Lens:	20mm f/1.1
Optional lenses:	Adaptor lens
Speed:	24 fps
Still picture:	No
Sound head:	Magnetic
Amplifier:	7W
Loudspeaker:	125mm, diameter
Power supply:	110-120V or 220-240V
Weight:	9.5kg
Dimensions:	280 x 220 x 360mm
Cost:	nia
Accessories:	Projection cabinets; carrying case
Other models:	1100ME, 1100M

8mm SPOOL SILENT PROJECTORS

These projectors are intended mainly for use in the domestic market. The original size of 8mm film is known as Standard 8 but this was superceded in 1965 by the Super 8 format which gives a bigger picture area. No Standard 8 only projectors are now available but a few machines are capable of projecting either format. Most of the projectors now available are of the automatic loading variety while some have a fast rewind with the film still loaded in the film path. This makes it possible to review sections of the film without reloading.

Make:	**Bauer**
Model:	T21
Film format:	Super 8, Standard 8, Single 8
Lamp:	8V, 50W
Lens:	16.5-30mm zoom, f/1.5
Threading:	Automatic
Reel size:	120m
Speed:	9 and 18 fps
Still picture:	No
Reverse:	No
Rewind:	nia
Power supply:	Mains
Weight:	nia
Dimensions:	nia
Cost:	(b)
Accessories:	None
Other models:	T49, similar to above with 12V, 100W lamp; 9, 12, 18 and 24 fps film speed

Make:	**Elmo**
Model:	K-100SM
Film format:	Super 8, Standard 8
Lamp:	12V, 100W
Lens:	15-25mm zoom, f/1.3
Threading:	Automatic
Reel size:	120m
Speed:	Variable 14-24 fps
Still picture:	Yes
Reverse:	Yes
Rewind:	Reel-to-Reel, fast
Power supply:	Mains
Weight:	6.5kg
Dimensions:	nia
Cost:	(a)
Accessories:	None
Notes:	Synchronous start with tape recorder possible

Make:	**Eumig**
Model:	604 Super 8
Film format:	Super 8
Lamp:	12V, 100W
Lens:	17-30mm zoom, f/1.4
Threading:	Automatic
Reel size:	120m
Speed:	18 fps
Still picture:	No
Reverse:	Yes
Rewind:	Through projector, fast
Power supply:	Mains
Weight:	6kg
Dimensions:	nia
Cost:	(a)
Accessories:	Cover; daylight viewer
Other models:	614D, as above plus Standard 8; still picture facilities; 6, 9 and 18 fps film speed

Make:	**Eumig**
Model:	624 D
Film format:	Super 8, Standard 8
Lamp:	12V, 100W
Lens:	15-30mm zoom, f/1.3
Threading:	Automatic
Reel size:	120m
Speed:	3, 6, 9, 12 and 18 fps
Still picture:	Yes
Reverse:	Yes
Rewind:	Through projector, fast

Power supply: Mains
Weight: 6kg
Dimensions: nia
Cost: (b)
Accessories: Dust cover; daylight viewer; cassette sync lead
Notes: Sync facility with cassette recorder

Make: **Eumig**
Model: R 2000
Film format: Super 8, Standard 8
Lamp: 12V, 100W
Lens: 7mm f/1.3
Threading: Automatic
Reel size: 120m
Speed: 3, 6, 9, 12 and 18 fps
Still picture: Yes
Reverse: Yes
Rewind: Through projector, fast
Power supply: Mains
Weight: 9.5kg
Dimensions: nia
Cost: (b)
Accessories: Cassette sync lead
Notes: Auto end stop; fast forward; built-in screen; front or rear projection; sync facility with cassette recorder

Make: **Raynox**
Model: 2000
Film format: Super 8, Standard 8
Lamp: 8V, 50W
Lens: 15-25mm zoom, f/1.5
Threading: Automatic
Reel size: 120m
Speed: 18 and 24 fps
Still picture: Yes
Reverse: Yes
Rewind: Fast
Power supply: Mains
Weight: nia
Dimensions: nia
Cost: (a)
Accessories: None
Other models: 3000, as above with 12V, 100W lamp (a)

8mm SOUND PROJECTORS

These cine projectors are only available in the Super 8 format when dealing with sound film, but a few are dual format and can project silent Standard 8 film. The sound playback facility consists of an amplifier and magnetic head which can decode the audio signal from a magnetic strip which is added to the film after editing. Some machines can also be used to record the sound track thus making it possible for the operator to produce the complete film, both visual and sound. The level of sophistication varies from machine to machine but at the top of the range it is possible to cope with stereo sound and mixing signals from several sources.

Make: **Bauer**
Model: T171
Film format: Super 8, Standard 8
Lamp: 12V, 100W
Lens: 16.5-30mm zoom
Threading: Automatic
Reel size: 180m
Speed: 18 and 24 fps
Still picture: No
Reverse: No
Rewind: Fast
Sound head: Magnetic
Amplifier: 6W
Loudspeaker: Integral
Power supply: Mains
Weight: nia
Dimensions: nia
Cost: (b)
Accessories: None
Other models: T179 – as above with sound on sound facility (c)

Make: **Bauer**
Model: TR 100
Film format: Super 8
Lamp: 12V, 100W
Lens: nia
Threading: Automatic
Reel size: 180m
Speed: 18 and 24 fps
Still picture: nia
Reverse: nia
Rewind: nia
Sound head: Magnetic
Amplifier: 10W
Loudspeaker: Integral
Power supply: Mains
Weight: nia
Dimensions: nia
Cost: (d)
Accessories: None
Notes: Integral back-projection screen; front-projection

Make: **Bauer**
Model: T 600
Film format: Super 8, Standard 8
Lamp: 15V, 120W
Lens: 16.5-30mm zoom, f/1.3
Threading: Automatic
Reel size: 240m
Speed: 18 and 24 fps
Still picture: Yes
Reverse: Yes
Rewind: Fast
Sound head: Magnetic
Amplifier: 20 plus 20W, stereo
Power supply: Mains
Weight: nia
Dimensions: nia
Cost: (g)
Accessories: None
Other models: T610, as above with sound on sound and microprocessor control

Make: **Beaulieau**
Model: 708 EL
Film format: Super 8, Standard 8
Lamp: 15V, 150W
Lens: 11-30mm zoom, f/1.1
Optional lenses: None
Threading: Automatic
Reel size: 700m
Speed: 18 and 24 fps
Still picture: No
Reverse: No

Rewind: Reel-to-reel, fast
Sound head: Magnetic
Amplifier: 24W; 50Hz to 12kHz; bass and treble tone controls
Loudspeaker: Integral
Power supply: 110/127/220/240V
Weight: nia
Dimensions: 370 x 232 x 315mm
Cost: (k)
Accessories: Mixing box
Other models: 708 EL stereo, as above with stereo sound; 708 EL stereo High Power Lamp, similar to above but Super 8 only with 250W high power lamp
Notes: Public address facility

Make: **Braun**
Model: Visacustic 2000 digital
Film format: Super 8, Standard 8
Lamp: 15V, 150W
Lens: 11-30mm zoom, f/1.1
Optional lenses: 40mm f/1.5
Threading: Automatic
Reel size: 240m
Speed: 16⅔, 18, 24, 25 fps
Still picture: Yes
Reverse: Yes
Rewind: Reel-to-reel, fast
Sound head: Magnetic
Amplifier: 2 x 15W, stereo
Loudspeaker: Integral
Power supply: 110/130/220/240V
Weight: 11.4kg
Dimensions: 289 x 152 x 425mm
Cost: (f)
Accessories: Ext speaker; remote control; dust cover

Make: **Elmo**
Model: ST 180M
Film format: Super 8
Lamp: 12V, 100W
Lens: 15-25mm zoom, f/1.3
Threading: Automatic
Reel size: 180m
Speed: 18 and 24 fps
Still picture: No
Reverse: Yes
Rewind: Reel-to-reel
Sound head: Magnetic
Amplifier: 5W
Loudspeaker: Integral
Power supply: Mains
Weight: 8.5kg
Dimensions: nia
Cost: (b)
Accessories: None
Other models: ST 180M-0, as above plus optical sound head. SC 18M, similar to ST 180M with built-in screen, front and rear projection (b). SC 18M-0, as SC 18M plus optical sound head

Make: **Eumig**
Model: ST 600D-M
Film format: Super 8
Lamp: 12V, 100W
Lens: 15-25mm zoom, f/1.3
Threading: Automatic
Reel size: 180m
Speed: 18 and 24 fps
Still picture: No
Reverse: Yes
Rewind: Reel-to-reel
Sound head: Magnetic
Amplifier: 6W
Loudspeaker: Integral
Power supply: Mains
Weight: 8.5kg
Dimensions: nia
Cost: nia
Accessories: None
Other models: ST 600D-M0, as above plus optical sound head

Make: **Eumig**
Model: ST 1200 HD-M
Film format: Super 8
Lamp: 15V, 150W
Lens: 15-25mm zoom, f/1.3
Threading: Automatic
Reel size: 360m
Speed: 18 and 24 fps
Still picture: No
Reverse: Yes
Rewind: Reel-to-reel
Sound head: Magnetic
Amplifier: 15W
Loudspeaker: Integral
Power supply: Mains
Weight: 9.7kg
Dimensions: nia
Cost: (e)
Accessories: None
Other models: ST 1200 HD-M0, as above plus optical sound head
Notes: 2 track sound system for sound-on-sound recording

Make: **Eumig**
Model: S932
Film format: Super 8
Lamp: 12V, 100W
Lens: 17-30mm zoom, f/1.4
Threading: Automatic
Reel size: 240m
Speed: 18 and 24 fps
Still picture: No
Reverse: Yes
Rewind: Reel-to-reel
Sound head: Magnetic
Amplifier: 5W
Loudspeaker: Integral
Power supply: Mains
Weight: 8.5kg
Dimensions: nia
Cost: (b)
Accessories: Dust cover
Other models: S934 – as above with 10W amplifier (b). S936 – as above with 20W amplifier; 15-30mm 200m, f/1.3 lens

Make: **Eumig**
Model: RS 3000
Film format: Super 8
Lamp: 12V, 100W
Lens: 13.6mm f/1.3
Threading: Automatic
Reel size: 180m
Speed: 18 and 24 fps
Still picture: No
Reverse: Yes
Rewind: Reel-to-reel
Sound head: Magnetic
Amplifier: 15W
Loudspeaker: Integral
Power supply: Mains

Weight:	11.5kg
Dimensions:	nia
Cost:	(c)
Accessories:	None
Notes:	Integral screen; front and rear projection

Make:	**Eumig**
Model:	GS 800-M
Film format:	Super 8
Lamp:	12V, 100W
Lens:	15-25mm zoom, f/1.3
Threading:	Automatic
Reel size:	240m
Speed:	18 and 24 fps
Still picture:	No
Reverse:	Yes
Rewind:	Reel-to-reel
Sound head:	Magnetic
Amplifier:	8 plus 8W, stereo
Loudspeaker:	Integral or extension
Power supply:	Mains
Weight:	9.4kg
Dimensions:	nia
Cost:	(d)
Accessories:	None
Other models:	GS 800-M0 – as above plus optical head
Notes:	Sound-on-sound recording possible

Make:	**Eumig**
Model:	S938
Film format:	Super 8
Lamp:	15V, 150W
Lens:	14-30mm zoom, f/1.3
Threading:	Automatic
Reel size:	240m
Speed:	18 and 24 fps
Still picture:	No
Reverse:	Yes
Rewind:	Reel-to-reel
Sound head:	Magnetic
Amplifier:	2 x 20W, stereo
Loudspeaker:	nia
Power supply:	Mains
Weight:	9kg
Dimensions:	nia
Cost:	(d)
Accessories:	None
Other models:	S940 – as above with 12.5-25mm zoom, f/1.2; microprocessor mixing and control unit (e)

Make:	**Eumig**
Model:	GS 1200
Film format:	Super 8
Lamp:	24V, 200W
Lens:	12.5-25mm zoom, f/1.1
Threading:	Automatic
Reel size:	360m
Speed:	18 and 24 fps
Still picture:	Yes
Reverse:	Yes
Rewind:	Reel-to-reel
Sound head:	Magnetic and optical
Amplifier:	15 plus 15W, stereo
Loudspeaker:	External
Power supply:	Mains
Weight:	16kg
Dimensions:	nia
Cost:	(g)
Accessories:	None
Notes:	Sound-on-sound recording possible; lamp economy switch; remote control; slow motion projection

Make:	**Fumeo**
Model:	9119
Film format:	Super 8
Lamp:	24V, 200W
Lens:	16.5-30mm 200m, f/1.3
Optional lenses:	See Fumeo 9914
Threading:	Automatic
Reel size:	750m
Speed:	18 and 24 fps
Still picture:	nia
Reverse:	Yes
Rewind:	Reel-to-reel, fast
Sound head:	Optical and magnetic
Amplifier:	25W; bass, treble and presence controls
Loudspeaker:	Integral
Power supply:	200-240V
Weight:	15.2kg
Dimensions:	320 x 420 x 210mm
Cost:	nia
Accessories:	Ext speaker; audio mixer unit
Other models:	9120 – as above, with recording facilities. 9139 as 9120, with arc lamp

Make:	**Fumeo**
Model:	9914
Film format:	Super 8
Lamp:	12V, 100W
Lens:	16.5-30mm zoom, f/1.3
Optional lenses:	11-30mm zoom, f/1.1; 15.5-28mm f/1.2; 50mm f/1.4
Threading:	Automatic
Reel size:	180m
Speed:	18 and 24 fps
Still picture:	No
Reverse:	No
Rewind:	Reel-to-reel, fast
Sound head:	Magnetic
Amplifier:	10W
Loudspeaker:	Integral
Power supply:	110-120/130 220-240V
Weight:	7.5kg
Dimensions:	230 x 290 x 180mm
Cost:	nia
Accessories:	Ext speaker
Other models:	9910 – as above with optical and magnetic sound heads

Make:	**Minolta**
Model:	Sound 7000
Film format:	Super 8, Standard 8
Lamp:	12V, 100W
Lens:	15-23mm zoom, f/1.3
Threading:	Automatic
Reel size:	180m
Speed:	18 and 24 fps
Still picture:	nia
Reverse:	nia
Rewind:	nia
Sound head:	Magnetic
Amplifier:	2W
Loudspeaker:	Integral
Power supply:	Mains
Weight:	nia
Dimensions:	nia
Cost:	(c)
Accessories:	None
Notes:	Sound-on-sound recording

Make:	**Sanyo**
Model:	SHV-2000
Film format:	Super 8, Standard 8
Lamp:	8V, 50W
Lens:	8mm f/1.4

Threading: Automatic
Reel size: 100m
Speed: 18 and 24 fps
Still picture: Yes
Reverse: No
Rewind: Fast
Sound head: Magnetic
Amplifier: 2W
Loudspeaker: Integral; 100mm diameter
Power supply: Mains
Weight: 9kg
Dimensions: 480 x 355 x 140mm
Cost: (c)
Accessories: None
Notes: Built-in screen, 276 x 212mm, for back projection; front projection

16mm FILM PROJECTORS

Most 16mm film projectors use low voltage halogen lamps, although a few are supplied with xenon arc lamps which give a very high light output but require bulky control equipment.

The lenses fitted, or available as optional extras, have two parameters:

1. *The focal length* (expressed in mm). The greater this is, the further from the screen the projector will be for the same image size. Zoom lenses have a continuously variable focal length, between two limits, and are useful when the projector is going to be used in a variety of situations with different projector to screen distances.
2. *The aperture* (f number). The smaller this number is, the greater the amount of light which is allowed through the lens.

Threading of the film is carried out in three main ways:

1. *Manual.* In this case the film is inserted around the sprockets, through the gate mechanism, round the sound head and up to the take-up spool by the operator.
2. *Automatic.* The film is inserted into an opening at the front of the projector and is then fed through without having to be handled by the operator. With automatic threading it is often difficult, or impossible, to thread or unthread the projector manually.
3. *Slot loading.* The film is inserted in a slot, or series of slots, in the side of the projector. This system can be regarded as having the advantages of automatic threading with few of its disadvantages.

The speed of the film through the projector varies depending on whether the film being projected was originally produced as a silent film – 18 frames per second (fps) – or as a sound film – 24 fps.

All projectors have a fast rewind facility to enable the film to be rewound after showing, most do this from reel-to-reel without passing through the mechanism but a few projectors can rewind rapidly without having to unthread the film.

The still picture control enables a single frame of the film to be projected and held stationary. However, the brightness is reduced because a heat filter is placed between the film and the lamp to prevent damage to the film by overheating.

Most 16mm sound tracks are of the optical variety but, since some films have a magnetic stripe for sound recording, a number of projectors have both optical and magnetic sound heads. A few projectors in this class permit the recording of the sound track by the operator.

Make: **Bell & Howell**
Model: Filmosound 1585 CX
Lamp: 24V, 250W
Lens: 50mm f/1.2
Optional lenses: 25mm f/1.6; 38mm f/1.5; 64mm f/1.5; 76mm f/1.6; 100m f/1.6; 32-65mm f/1.3 zoom
Threading: Automatic
Reel size: 120m supplied
Speed: 24 fps
Still picture: No
Reverse: Yes
Rewind: Reel-to-reel
Remote control: No
Sound head: Optical
Amplifier: 20W
Loudspeaker: Integral
Power supply: 120V, 220V plus 250V
Weight: 16kg
Dimensions: 340 x 340 x 290mm
Cost: (g)
Accessories: None
Other models: 1592 C5X – as above with still picture and detachable speaker (h)

Make: **Bell & Howell**
Model: Filmosound 1680 GS
Lamp: 24V, 250W
Lens: 50mm f/1.2
Optional lenses: See Filmosound 1585 CX
Threading: Slot loading
Reel size: nia
Speed: 18 and 24 fps
Still picture: Yes
Reverse: Yes
Rewind: Reel-to-reel
Remote control: No
Sound head: Optical
Amplifier: 20W
Loudspeaker: Integral and detachable
Power supply: 100 to 250V
Weight: 15kg
Dimensions: 402 x 365 x 256mm
Cost: (h)
Accessories: None
Notes: Socket for microphone

Make: **Bell & Howell**
Model: TQ111-1692
Lamp: 24V, 250W
Lens: 50mm f/1.2
Optional lenses: See Filmosound 1585 CX
Threading: Automatic
Reel size: 490m supplied; 600m max
Speed: 18 and 24 fps
Still picture: No
Reverse: Yes
Rewind: Reel-to-reel
Remote control: No
Sound head: Optical
Amplifier: 12W
Loudspeaker: Integral; external socket

Power supply: 100 to 250V
Weight: 14.5kg
Dimensions: 402 x 365 x 260mm
Cost: (i)
Accessories: Plug in magnetic sound heads
Other models: TQ111 – 1693, as above with optical/magnetic sound heads (i)
Notes: Lamp economy switch

Make: **Bell & Howell**
Model: TQ111 Specialist – 1695
Lamp: 24V, 250W
Lens: 50mm f/1.2
Optional lenses: See Filmosound 1585 CX
Threading: Automatic
Reel size: 490m supplied; 600m max
Speed: 18 and 24 fps
Still picture: Yes
Reverse: Yes
Rewind: Reel-to-reel
Remote control: Socket provided
Sound head: Optical
Amplifier: 20W
Loudspeaker: Integral plus detachable speaker
Power supply: 100-250V
Weight: 16.3kg
Dimensions: 402 x 303 x 365mm
Cost: (i)
Accessories: Remote control unit
Other models: 1698 – as above with optical/magnetic soundlead (k). 1694 – as 1698 with magnetic recording facility (k).
Notes: Animation facility; bass and treble tone controls

Make: **Bell & Howell**
Model: 1568
Lamp: 300W, gas discharge
Lens: 50mm f/1.2
Optional lenses: See Filmosound 1585 CX
Threading: Automatic
Reel size: 120m to 600m
Speed: 24 fps
Still picture: No
Reverse: Yes
Rewind: Reel-to-reel
Remote control: Yes
Sound head: Optical
Amplifier: 20W
Loudspeaker: 2, detachable
Power supply: 105-132V, 220-240V
Weight: 17.5kg
Dimensions: 405 x 360 x 305mm
Cost: (k)
Accessories: None
Notes: Microphone input

Make: **Bolex**
Model: 501
Lamp: 24V, 250W
Lens: 50mm f/1.3
Optional lenses: 70mm f/1.6; 35-65mm f/1.6 zoom
Threading: Automatic
Reel size: 600m
Speed: 18 and 24 fps
Still picture: No
Reverse: Yes
Rewind: Reel-to-reel
Remote control: No
Sound head: Optical
Amplifier: 20W
Loudspeaker: Detachable
Power supply: 118/127/220/240V
Weight: 17kg
Dimensions: 480 x 380 x 240mm
Cost: (i)
Accessories: None
Other models: 510 – as above with optical and magnetic sound head, still picture, continuous adjustment of film speed from 12 to 26 fps (k)
Notes: Microphone and pick-up inputs

Make: **Bolex**
Model: 521
Lamp: 24V, 250W
Lens: 50mm f/1.3
Optional lenses: See 501
Threading: Automatic
Reel size: 600m
Speed: Adjustable 12 to 26 fps
Still picture: Yes
Reverse: Yes
Rewind: Reel-to-reel
Remote control: Socket provided
Sound head: Optical/magnetic
Amplifier: 25W
Loudspeaker: Integral/monitor; detachable main speaker
Power supply: 118/127/220/240V
Weight: 18kg
Dimensions: 480 x 380 x 240mm
Cost: (k)
Accessories: Remote control unit; widescreen attachment
Notes: Magnetic recording facility

Make: **Elf**
Model: SLO
Lamp: 24V, 250W
Lens: 50mm f/1.2
Optional lenses: 25mm f/1.5; 38mm f/1.5; 63mm f/1.6; 75mm f/2; 88mm f/2; 100mm f/2.2; 35-65mm f/1.6
Threading: Manual
Reel size: 600m
Speed: 18 and 24 fps
Still picture: No
Reverse: No
Rewind: Reel-to-reel or through film path
Remote control: No
Sound head: Optical
Amplifier: 15W
Loudspeaker: Integral
Power supply: 240V
Weight: 14.1kg
Dimensions: 380 x 350 x 290mm
Cost: (i)
Accessories: None
Other models: SL1 – as above with additional detachable speaker (i). SL2 – as SL1 with optical/magnetic sound head (i). SL02 – as SL0 with optical/magnetic sound head (i)

Make: **Elf**
Model: NTO
Lamp: 24V, 250 or 200W
Lens: 50mm f/1.2
Optional lenses: See SLO
Threading: Automatic or manual
Reel size: 600m
Speed: 18 and 24 fps
Still picture: Yes
Reverse: Yes
Rewind: Reel-to-reel
Remote control: No
Sound head: Optical
Amplifier: 15W
Loudspeaker: Integral

Power supply: 240V
Weight: 16kg
Dimensions: 350 x 350 x 290mm
Cost: (i)
Accessories: None
Other models: NT1 – as above with additional detachable speaker (i). NT2 – as NT1 with optical/magnetic sound head (j). NT3 – as NT2 with magnetic recording facility (k). NT1RC – as NT1 with cordless remote control (k). NT2RC – as NT2 with cordless remote control (k)

Make: **Elmo**
Model: 16-CL
Lamp: 24V, 250W
Lens: 50mm f/1.2
Optional lenses: 12.5mm f/1.8; 20mm f/1.4; 75mm f/1.8; 50-100mm f/1.7 zoom
Threading: Semi-automatic
Reel size: 600m
Speed: 24 fps
Still picture: No
Reverse: No
Rewind: Reel-to-reel or through film path
Remote control: No
Sound head: Optical/magnetic
Amplifier: 10W
Loudspeaker: Integral; external socket
Power supply: AC single phase
Weight: 14kg
Dimensions: 354 x 265 x 220mm
Cost: (g)
Accessories: Extension speaker; rear projection screen and mirror; zoom converter; cinemascope lens

Make: **Elmo**
Model: 16AA
Lamp: 24V, 250W
Lens: 50mm f/1.2
Optional lenses: See 16 CL
Threading: Automatic
Reel size: 360m standard (600m max)
Speed: 24 fps (18 fps available)
Still picture: Yes
Reverse: Yes
Rewind: Reel-to-reel
Remote control: Yes
Sound head: Optical/magnetic
Amplifier: 25W
Loudspeaker: Two integral; external socket
Power supply: AC single phase
Weight: 16kg
Dimensions: 410 x 340 x 250mm
Cost: (i)
Accessories: See 16CL
Other models: 16AAR – as above with magnetic recording facility (k)

Make: **Elmo**
Model: 16F
Lamp: 24V, 250W
Lens: 50mm f/1.3
Optional lenses: See 16CL
Threading: Manual
Reel size: 480m standard, 600m max
Speed: 24 fps (18 fps available)
Still picture: Yes
Reverse: Yes
Rewind: Reel-to-reel
Remote control: Socket provided
Sound head: Optical/magnetic

Amplifier: 15W continuous, 20W max
Loudspeaker: Integral
Power supply: AC single phase
Weight: 19kg
Dimensions: 470 x 390 x 260mm
Cost: (i)
Accessories: See 16CL
Other models: 16 FS – as above with optical sound head only. 16 FR – as 16F with magnetic recording facility (j)

Make: **Elmo**
Model: 16CLX
Lamp: 250W xenon-arc lamp
Lens: 50mm f/1.2
Optional lenses: See 16CL
Threading: Semi-automatic
Reel size: 600m
Speed: 24 fps
Still picture: No
Reverse: No
Rewind: Reel-to-reel or through film path
Remote control: No
Sound head: Optical/magnetic
Amplifier: 25W
Loudspeaker: Integral; external socket
Power supply: AC single phase
Weight: 17.5kg
Dimensions: 350 x 290 x 230mm
Cost: (k)
Accessories: See 16CL

Make: **Fumeo**
Model: 9250

Lamp: 24V, 200W
Lens: 50mm f/1.3
Optional lenses: See 9270
Threading: Manual
Reel size: 750m
Speed: 18 and 24 fps
Still picture: No
Reverse: No
Rewind: nia
Remote control: No
Sound head: Optical
Amplifier: 15W (25W max)
Loudspeaker: Integral 6W; external socket
Power supply: 200-250V
Weight: 14.5kg
Dimensions: 390 x 290 x 170mm
Cost: (f)
Accessories: See 9270
Other models: 9251 – as above with optical/magnetic sound head (g)
Notes: Separate bass and treble controls; microphone/pick-up input

Make: **Fumeo**
Model: 9226
Lamp: 24V, 250W
Lens: 50mm f/1.3
Optional lenses: See 9270
Threading: Manual
Reel size: 750m
Speed: 18 and 24 fps
Still picture: Yes
Reverse: Yes
Rewind: nia
Remote control: No
Sound head: Optical
Amplifier: 25W (35W max)
Loudspeaker: Integral 6W; external socket
Power supply: 200-250V
Weight: 14.5kg
Dimensions: 390 x 290 x 170mm
Cost: (g)
Accessories: See 9270
Other models: 9227 – as above with optical/magnetic sound head (h)
Notes: See 9260

Make: **Fumeo**
Model: 9260
Lamp: 24V, 200W
Lens: 50mm f/1.3
Optional lenses: See 9270
Threading: Manual
Reel size: 750m
Speed: 18 and 24 fps
Still picture: No
Reverse: No
Rewind: nia
Remote control: No
Sound head: Optical
Amplifier: 25W (35W max)
Loudspeaker: Integral 6W; external socket
Power supply: 200-250V
Weight: 14.5kg
Dimensions: 390 x 290 x 170mm
Cost: (g)
Accessories: See 9270
Other models: 9261 – as above with optical/magnetic sound head (h)
Notes: Separate bass and treble controls; microphone and pick-up inputs with independent mixing controls

Make: **Fumeo**
Model: 9270
Lamp: 24V, 250W
Lens: 50mm f/1.3
Optional lenses: 35mm; 40mm; 65mm; 75mm; 35-65mm zoom
Threading: Manual
Reel size: 750m
Speed: 18 and 24 fps
Still picture: No
Reverse: No
Rewind: nia
Remote control: No
Sound head: Optical
Amplifier: 25W (40W max)
Loudspeaker: Integral 6W; external socket
Power supply: 200-250V
Weight: 14.5kg
Dimensions: 390 x 290 x 170mm
Cost: (h)
Accessories: Removable arms for 1500m reels; change over unit; extension loudspeakers
Other models: 9271 – as above with optical/magnetic sound head (i)
Notes: Separate bass and treble controls and presence filter

Make: **Fumeo**
Model: 9272
Lamp: 24V, 250W
Lens: 50mm f/1.3
Optional lenses: 35mm; 40mm; 65mm; 75mm; 35-65mm zoom
Threading: Manual
Reel size: 750m
Speed: 18 and 24 fps
Still picture: No
Reverse: No
Rewind: nia
Remote control: No
Sound head: Optical
Amplifier: 25W (40W max)
Loudspeaker: Integral 6W; external socket
Power supply: 200-250V
Weight: 14.5kg
Dimensions: 390 x 290 x 170mm
Cost: (h)
Accessories: Removable arm for 1500m reels; change over unit; extension loudspeakers
Other models: 9271 – as above with optical/magnetic sound head (i)
Notes: Separate bass and treble controls and presence filter

Make: **Hokushin**
Model: SC10
Lamp: 24V, 250W
Lens: 50mm f/1.2
Optional lenses: Zoom anamorphic
Threading: Automatic
Reel size: 600m max
Speed: 24 fps
Still picture: No
Reverse: No
Rewind: High speed, no reel change
Remote control: Accessory
Sound head: Optical
Amplifier: 15W (undistorted) 25W (max)
Loudspeaker: Integral
Power supply: 100-130V/200-250V
Weight: 15kg
Dimensions: 380 x 364 x 250mm

Cost: (g)
Accessories: Recording adaptor for SC10M and SC10MF
Other models: SC10F – as above with detachable twin speaker unit (h). SC10M – as SC10 with optical and magnetic sound head; still projection facility (h). SC10MF – as SC10M with detachable twin speaker unit (h). SC10S – as SC10M with synchronous motor, suitable for televising

Make: **Hokushin**
Model: SC11
Lamp: 24V, 250W
Lens: 50mm f/1.2
Optional lenses: Zoom anamorphic
Threading: Manual
Reel size: 600m (max)
Speed: 24 fps
Still picture: No
Reverse: No
Rewind: High speed, no reel change
Remote control: Accessory
Sound head: Optical
Amplifier: 15W (undistorted); 25W (max)
Loudspeaker: Integral
Power supply: 100-130V/200-250V
Weight: 15kg
Dimensions: 380 x 364 x 250mm
Cost: (g)
Accessories: Recording adaptor for SC11M and SC11MF
Other models: SC11F – as above with detachable twin speaker unit (g). SC11M – as SC11F with optical and magnetic sound head; still projection facility (h). SC11MF – as SC11M with detachable twin speaker unit (h)

Make: **Microtecnia**
Model: Micron 27J
Lamp: 24V, 250W
Lens: 50mm f/1.3
Optional lenses: 35mm f/1.3; 60mm f/1.5; 75mm f/1.7; 35-60mm f/1.5 zoom
Threading: manual
Reel size: nia
Speed: 18 and 24 fps
Still picture: No
Reverse: Yes
Rewind: Reel-to-reel
Remote control: No
Sound head: Optical
Amplifier: 22W
Loudspeaker: Detachable
Power supply: nia
Weight: nia
Dimensions: 470 x 417 x 210mm
Cost: (g)
Accessories: None
Other models: 27JA – as above with automatic threading (h). 27JM – as 27J with optical/magnetic sound head (i). 27JAM – as 27JM with automatic threading (i)

Make: **Microtecnia**
Model: Micron 28J
Lamp: 24V, 250W
Lens: 50mm f/1.3
Optional lenses: See Micron 27J
Threading: Manual
Reel size: nia
Speed: 18 and 24 fps; 15 to 26 fps continuous
Still picture: Yes
Reverse: Yes
Rewind: Reel-to-reel
Remote control: No
Sound head: Optical
Amplifier: 22W
Loudspeaker: Detachable
Power supply: nia
Weight: nia
Dimensions: 470 x 417 x 210mm
Cost: (i)
Accessories: None
Other models: 28JA – as above with automatic threading (i). 28JM – as 28J with optical/magnetic sound head (j). 28JAM – as 28JM with automatic threading (j)

Make: **Rank Aldis**
Model: Series 4-0254
Lamp: 24V, 250W
Lens: 50mm f/1.2
Optional lenses: 25mm f/1.4; 35mm f/1.3; 65mm f/1.5; 75mm f/1.6; 35-65mm f/1.6 zoom
Threading: Automatic or manual
Reel size: nia
Speed: 18 and 24 fps
Still picture: Yes
Reverse: Yes
Rewind: Reel-to-reel
Remote control: No
Sound head: Optical
Amplifier: 25W
Loudspeaker: Integral
Power supply: 110/130/220/240/250V
Weight: 90kg
Dimensions: 392 x 322 x 226mm
Cost: (i)
Accessories: Extension speaker
Other models: Series 4-M254 – as above with optical/magnetic sound head (j). Series 4-MR254 – as M254 with magnetic recording facility (k)
Notes: Microphone and pick-up inputs; bass and treble tone controls; economy lamp switch

Make: **Singer**
Model: Instaload 2121
Lamp: 24V, 200W
Lens: 50mm f/1.4
Optional lenses: 25mm f/1.9; 40mm f/1.6; 65mm f/1.6
Threading: Manual
Reel size: nia
Speed: 24 fps
Still picture: Yes
Reverse: Yes
Rewind: Reel-to-reel or through film path
Remote control: No
Sound head: Optical
Amplifier: 15W
Loudspeaker: Integral
Power supply: 110/130/220/240/250V
Weight: 18kg
Dimensions: 400 x 320 x 280mm
Cost: (h)
Accessories: Extension speaker
Other models: Instaload 2171 – as above with fast forward and footage counter (h)
Notes: Economy lamp switch

Make: **Viewlex**
Model: System-16
Lamp: nia

Lens: nia
Optional lenses: nia
Threading: Automatic/manual
Reel size: nia
Speed: 6 or 24 fps
Still picture: Yes
Reverse: Yes
Rewind: Yes
Remote control: All facilities
Sound head: Optical
Amplifier: 20W
Loudspeaker: Integral; external socket
Power supply: 100-240V
Weight: nia
Dimensions: nia
Cost: (f)
Accessories: None
Notes: Modular construction facilitates rapid repair

AUDIO

AMPLIFIED LOUDSPEAKERS

An amplified loudspeaker consists of a power amplifier and loudspeaker in the same cabinet. The unit will not have a signal source built-in, but will accept the signal from some external source such as a cassette deck, radio or record deck. It is intended to amplify the signal to a level where it can be used for larger audiences. The important parameters are 'power output', which must be adequate for the size of auditorium being used, and 'frequency range', which must be sufficient for the type of sound being reproduced. Music requires a much wider frequency range than speech.

Make: **Coomber**
Model: 209
Output power: 15W
Loudspeaker: 254mm, diameter
Frequency range: nia
Tone controls: Bass and treble
Inputs: Tape
Outputs: Ext speaker
Power supply: 240V
Weight: 5.2kg
Dimensions: 305 x 470 x 150mm
Cost: (a)
Accessories: None
Other models: 210 – as above with microphone input for public address (a)

Make: **Goodsell**
Model: LAS 12
Output power: 12W
Loudspeaker: 200mm, diameter
Frequency range: 50Hz-20kHz
Tone controls: Bass and treble
Inputs: Mic; line
Outputs: Line; ext speaker
Power supply: 240V
Weight: 9kg
Dimensions: 300 x 280 x 450mm
Cost: (a)
Accessories: None
Other models: LAS 20, as above with 20W output

Make: **Goodsell**
Model: SAS
Output power: 10W
Loudspeaker: 250mm, diameter
Frequency range: 50Hz-20kHz
Tone controls: Bass and treble
Inputs: Line
Outputs: None
Power supply: 220-240V
Weight: 5.4kg
Dimensions: 381 x 381 x 222mm
Cost: (a)
Accessories: None

Make: **TOA**
Model: WA 24
Output power: 7W
Loudspeaker: nia
Frequency range: 200Hz-10kHz
Tone controls: Speech/music switch
Inputs: Mic; line
Outputs: Line
Power supply: 240V AC or batteries
Weight: 6.5kg
Dimensions: 290 x 418 x 142mm
Cost: (b)
Accessories: Microphone

Make: **TOA**
Model: MA 007
Output power: 12W
Loudspeaker: 200mm, diameter
Frequency range: 100Hz-10kHz
Tone controls: Yes
Inputs: Mic; line
Outputs: Line; ext speaker
Power supply: 240V AC or batteries
Weight: 10.4kg
Dimensions: 340 x 500 x 225mm
Cost: (c)
Accessories: None
Other models: MA 007/C – as above with cassette unit (d). MA 007/8t – as above with 8 track unit (d)
Notes: Built-in electronic echo system

Make: **TOA**
Model: MA 220
Output power: 15W
Loudspeaker: 4 elliptical units
Frequency range: 100Hz-10kHz
Tone controls: Speech/music switch
Inputs: 2-mic; line
Outputs: Line; ext speaker
Power supply: 240V AC or 10-16V DC
Weight: 16.5kg
Dimensions: Height 1710mm

Cost: (d)
Accessories: None
Notes: Simultaneous use of 2 microphones and line input possible; parallel connection of more than 2 amplifiers possible

AUDIO DISTRIBUTION UNITS

These units are meant to enable more than one user to listen to the same audio programme using individual headphones, and make it possible for tapes etc to be listened to in the classroom situation without disturbing the surrounding students in other groups. The unit is usually plugged into the extension speaker outlet of the recorder or radio being used and do not require a power source of their own. Two basic types are produced:

1. Each individual headset is of the electro-mechanical type and the distribution unit divides the electrical output from the source so that it is equally spread among the separate headsets. This type often has volume control and/or switches for each student.
2. In the other variety the students are connected to a small central loudspeaker by hollow plastic tubes each ending in a stethescope type of earpiece. In this acoustic type of unit separate volume controls are not provided.

Make: **S G Brown**
Model: Primary Audio Set
Description: Small loudspeaker in metal box which can be acoustically connected to up to 6 stethescope type leadsets. Extension boxes available to increase number to 24.
Cost: (a)

Make: **Hanimex**
Model: Hanimex Listening Centre
Description: 2 x 4 headphone set with distribution box in carrying case. 1 programme can be fed to each set of 4 phones, or 1 programme to all 8. Individual volume controls; reversing switch to change programmes to each group
Cost: nia

Make: **Specialist AV**
Model: Listening Centres
Description: Direct distribution box fitted with 6.3mm jack sockets, 2, 4, 6 or 8 available. Alternatively 6 sockets each with individual volume controls. Supplied with or without headphones. Flying lead loudspeaker connection
Cost: nia

CARD READERS

Card readers operate using a card which has a magnetic backing or strip at the bottom of the card. The card can have writing or diagrams on the upper section while relevant audio information can be recorded on the magnetic section. As the card passes through the reader the student can hear the information on the audio track, either through headphones or a built-in loudspeaker, while also seeing the graphic information. Most of the machines have two audio tracks which enables the teacher to record on one track while the student can subsequently record his attempt on the second.

The length of the recording is limited in the case of cards which are pulled through the reader by the length of card. This usually means that about 8 seconds is the maximum recording time for cards with the magnetic strip. Generally, the readers which permit separate recording of the teacher's voice have a key, or hidden control, to make it impossible for the student to erase the teacher's track.

The majority of machines have a volume control and automatic recording level circuits. Some have a repeat key which quickly reverses the direction of the card to enable a section to be listened to again without having to re-insert the card.

Make: **Bell and Howell**
Model: Language Master 1727B
Functions: (i) Playback of teacher's voice; (ii) recording of pupil's voice; (iii) playback of pupils's voice. No facility for recording teacher's voice
Card speed: 5.71cm/s
Power supply: Battery
Weight: 1.3kg
Dimensions: 240 x 155 x 68mm
Cost: nia
Accessories: Audio-active headset; mains adaptor; dual headphone adaptor; 4 way junction box; dust cover; head cleaning strip
Notes: Socket for headset

Make: **Bell and Howell**
Model: Language Master 1757
Functions: (i) Record and playback teacher's voice; (ii) record and playback pupil's voice
Card speed: 5.71cm/s or 2.85cm/s
Power supply: 110/220/240V
Weight: 3.4kg
Dimensions: 306 x 230 x 160mm
Cost: (b)
Accessories: Audio-active headset; dual headphone adaptor; 4 way junction box; dust cover; head cleaning strip
Notes: Socket for headset; fast repeat key

Make: **Rank Aldis**
Model: 810
Functions: (i) Playback of teacher's voice; (ii) record/playback of pupil's voice;
Card speed: nia
Power supply: Mains/battery
Weight: nia
Dimensions: nia
Cost: (b)
Accessories: Headset; blank cards; programmes
Other models: 800 – as above; mains only; playback/record teacher's voice (b)

CASSETTE PLAYERS

A cassette player has no record facility and is only intended for the replay of pre-recorded tapes. It is especially useful in situations where students will only be required to listen to tapes since there is no possibility of accidental erasure of the recorded signal.

Make:	**Clarke and Smith**
Model:	1088/C/1
Output power:	1W
Loudspeaker:	130mm, circular
Frequency range:	nia
Tone controls:	Treble cut
Outputs:	Headphones
Pause:	No
Cue and review:	No
Tape counter:	No
Power supply:	240V
Weight:	2kg
Dimensions:	200 x 180 x 170mm
Cost:	(a)
Accessories:	None

Make:	**Coomber**
Model:	282
Output power:	1.2W
Loudspeaker:	127 x 76mm, elliptical
Frequency range:	nia
Tone controls:	No
Outputs:	Ext speaker; 8 headphone jacks
Pause:	No
Cue and review:	No
Tape counter:	3 digit
Power supply:	240V
Weight:	3.5kg
Dimensions:	140 x 219 x 340mm
Cost:	(a)
Accessories:	None

Make:	**Goodsell**
Model:	PCP 1285
Output power:	12W
Loudspeaker:	200 x 125mm, elliptical
Frequency range:	100Hz-80kHz
Tone controls:	Base and treble
Outputs:	Ext speaker
Pause:	Yes
Cue and review:	No
Tape counter:	3 digit
Power supply:	240V
Weight:	4.5kg
Dimensions:	330 x 300 x 180mm
Cost:	(b)
Accessories:	Dust cover
Other models:	PCP 2010 – as above with 20W output and 250mm circular speaker (b)
Notes:	Memory counter can be fitted

Make:	**Neal**
Model:	104
Output power:	Line only
Loudspeaker:	No
Frequency range:	35Hz-15kHz
Tone controls:	No
Outputs:	Line; headphones
Pause:	No
Cue and review:	No
Tape counter:	3 digit
Power supply:	nia
Weight:	6.4kg
Dimensions:	356 x 243 x 127mm
Cost:	nia
Accessories:	None

Make:	**Rank Aldis**
Model:	240
Output power:	740mW
Loudspeaker:	75mm, circular
Frequency range:	100Hz-8kHz
Tone controls:	Switched
Outputs:	Headphones
Pause:	No
Cue and review:	Yes
Tape counter:	3 digit
Power supply:	240V AC or 4x1.5 DC
Weight:	1.8kg
Dimensions:	229 x 216 x 83mm
Cost:	(a)
Accessories:	Headphones

Make:	**Telex (Avcom Systems Ltd)**
Model:	C 100
Output power:	10W
Loudspeaker:	Integral
Frequency range:	40Hz-10kHz
Tone controls:	Yes
Outputs:	nia
Pause:	Yes
Cue and review:	Yes
Tape counter:	Yes
Power supply:	220V
Weight:	6.8kg
Dimensions:	133 x 254 x 374mm
Cost:	(b)
Accessories:	Foot switch; headphones

CASSETTE RECORDERS

In a cassette recorder the tape is reduced in width to ⅛ inch and is completely enclosed in a plastic cassette which contains both the feed and take-up spools plus a pressure pad and lubricating system. As a result the tape itself does not need to be handled in any way and is automatically placed in contact with the erase and record/playback heads in the recorder. The tape speed is fixed at 2.4cm/s, although a few machines have a facility for altering this slightly to enable the pitch of the recorded signal to be varied slightly. The recording level in the recorder can be controlled in two ways, manually using a recording level control and adjusting this in conjunction with readings on a meter or automatically by switching in special circuitry which senses variations in the recording level and compensates for these. Some recorders can be controlled either way. Built-in microphones are becoming more evident on many recorders but generally these do not produce such good recordings as using the same machine with an external microphone.

The cue and review facility on some machines makes it possible to listen to the signal on the tape while it is being rapidly wound forward or back. This allows rapid spotting of certain points in the tape once the necessary ability to decode high speed sound has been acquired.

Make: **Aiwa**
Model: TM-400
Output power: nia
Loudspeaker: 77mm, circular
Frequency range: 50Hz-10kHz
Tone controls: No
Recording level: Automatic
Microphone: Built-in
Monitor on record: No
Inputs: Mic
Outputs: Headset
Pause: Yes
Cue and review: Yes
Tape counter: 3 digit
Power supply: 110-120V/220-240V AC, 6V DC
Weight: 1.7kg
Dimensions: 156 x 77 x 276mm
Cost: nia
Accessories: Headset HP-55
Notes: A 4 track 2 channel recorder, language lab facility enables one track to be repeatedly erased and re-recorded while the other, model track is unaltered

Make: **Bell and Howell**
Model: 3185
Output power: 2W
Loudspeaker: 127 x 76mm, elliptical
Frequency range: 100Hz-10kHz
Tone controls: Treble cut
Recording level: Automatic/manual (metered)
Microphone: Built-in
Monitor on record: No
Inputs: Mic; line
Outputs: Ext speaker; 4 headphone sockets
Pause: Yes
Cue and review: Yes
Tape counter: 3 digit
Power supply: 220-240V AC or batteries
Weight: 3.18kg
Dimensions: 262 x 287 x 87mm
Cost: (a)
Accessories: Microphone; headphones
Other models: 3196 – as above with 3W output, treble boost/cut, 6 headphone sockets (a)

Make: **Coomber**
Model: 307
Output power: 4.5W
Loudspeaker: 200 x 75mm, elliptical
Frequency range: nia
Tone controls: Bass and treble
Recording level: Automatic
Microphone: External
Monitor on record: nia
Inputs: Mic; line record; line amplify
Outputs: Line; ext speaker
Pause: Yes
Cue and review: No
Tape counter: 3 digit
Power supply: Mains
Weight: 5.7kg
Dimensions: 178 x 362 x 328mm
Cost: (b)
Accessories: Microphones; headphones; ext speaker
Other models: 309 – as above with 15W output (b)

Make: **Coomber**
Model: 341
Output power: 4.5W
Loudspeaker: 245mm, circular
Frequency range: nia
Tone controls: Bass and treble
Recording level: Automatic
Microphone: External
Monitor on record: nia
Inputs: Mic; line record; line amplify
Outputs: Line; ext speaker
Pause: Yes
Cue and review: No
Tape counter: 3 digit
Power supply: Mains
Weight: 6.0kg
Dimensions: 380 x 377 x 210mm
Cost: (b)
Accessories: Microphone; headphones; ext speaker
Other models: 343 – as above with 15W output (b)

Make: **Farnell**
Model: CC 15
Output power: 10W
Loudspeaker: Elliptical
Frequency range: nia
Tone controls: Bass and treble
Recording level: nia
Microphone: nia
Monitor on record: No
Inputs: Mic; line
Outputs: Ext speaker
Pause: nia
Cue and review: Yes
Tape counter: nia
Power supply: Mains
Weight: nia
Dimensions: nia
Cost: (a)
Accessories: Ext speaker
Notes: Mono/stereo recorder

Make: **Goodsell**
Model: CTR 585
Output power: 5W
Loudspeaker: 200 x 125mm, elliptical
Frequency range: Amplifier 50Hz-20kHz, tape deck 100Hz-8kHz
Tone controls: Bass and treble
Recording level: Automatic/manual (metered)
Microphone: Built-in or external
Monitor on record: Yes
Inputs: Mic; line
Outputs: Line; ext speaker
Pause: Yes
Cue and review: No
Tape counter: 3 digit
Power supply: 240V
Weight: 4.5kg
Dimensions: 330 x 300 x 180mm
Cost: (b)
Accessories: Dust cover; microphones; various connecting leads
Other models: CTR 510 – as above with 250mm circular loudspeaker (b). CTR 1285 – as above with 12W output (b)
Notes: Memory counter can be fitted

Make: **Grundig**
Model: CR 355
Output power: 1W
Loudspeaker: Integral
Frequency range: nia
Tone controls: No
Recording level: Automatic
Microphone: Built-in
Monitor on record: Yes
Inputs: Mic
Outputs: Ext speaker; line

Pause: Yes
Cue and review: No
Tape counter: No
Power supply: Mains/battery
Weight: 1.9kg
Dimensions: 210 x 80 x 280mm
Cost: (a)
Accessories: None
Other models: CR 455 – as above with tone control; 3 digit counter; record level meter (a). CR 485 – as CR 455 with 1.5W output and stereo facility (a)

Make: **ITT**
Model: SL 59
Output power: 1W
Loudspeaker: 100 x 70mm, elliptical
Frequency range: nia
Tone controls: No
Recording level: Automatic
Microphone: Built-in
Monitor on record: No
Inputs: Mic/line
Outputs: Line
Pause: No
Cue and review: No
Tape counter: No
Power supply: 240V AC or batteries
Weight: 1.8kg
Dimensions: 185 x 70 x 245mm
Cost: nia
Accessories: Microphone

Make: **ITT**
Model: ST 66
Output power: 1W
Loudspeaker: 100 x 70mm, elliptical
Frequency range: nia
Tone controls: Treble cut
Recording level: Automatic/manual (metered)
Microphone: Built-in
Monitor on record: No
Inputs: Mic/line
Outputs: Line; ext speaker
Pause: Yes
Cue and review: No
Tape counter: 3 digit
Power supply: 240V AC and batteries
Weight: 1.8kg
Dimensions: 185 x 70 x 245mm
Cost: nia
Accessories: Microphone

Make: **Marantz**
Model: C-170
Output power: nia
Loudspeaker: 90mm, circular
Frequency range: nia
Tone controls: No
Recording level: Automatic
Microphone: Built-in
Monitor on record: No
Inputs: Mic; line; remote
Outputs: Headphones; ext speaker
Pause: Yes
Cue and review: No
Tape counter: No
Power supply: 240V AC or batteries
Weight: nia
Dimensions: nia
Cost: (a)
Accessories: None
Other models: C-180 – as above with 3 digit counter; record level indicator; cue and review; tone control (a). C-190 – as C-180 with manual record level (metered); vari-speed pitch control (a)

Make: **Marantz**
Model: C-204
Output power: nia
Loudspeaker: 100mm, circular
Frequency range: nia
Tone controls: Yes
Recording level: Automatic/manual (metered)
Microphone: Built-in
Monitor on record: Yes
Inputs: Mic; line; remote control
Outputs: Line; headphones; ext speaker
Pause: Yes
Cue and review: Yes
Tape counter: 3 digit
Power supply: Mains/battery
Weight: 3.1kg
Dimensions: 300 x 195 x 82mm
Cost: (b)
Accessories: Microphone; telephone pick-up
Other models: C-205 – as above with separate record and playback heads; memory rewind/repeat; bias and equalisation adjustable; separate record and playback volume controls (b)
Notes: Vari-speed pitch control

Make: **Marantz**
Model: CD-320
Output power: 1.2W
Loudspeaker: 100mm, circular
Frequency range: 40Hz-14kHz
Tone controls: Yes
Recording level: Automatic/manual (metered)
Microphone: No
Monitor on record: Yes
Inputs: Mic; line; telephone pick-up
Outputs: Line; headphones; ext speaker
Pause: Yes
Cue and review: Yes
Tape counter: 3 digit
Power supply: 220-240V AC, 6V DC, batteries or rechargeable battery pack
Weight: 3.2kg
Dimensions: 293 x 82 x 195mm
Cost: (b)
Accessories: Battery pack; telephone pick-up
Other models: CD-330 – as above with separate record and playback heads (b)
Notes: Stereo recorder

Make: **National Panasonic**
Model: RQ-2735
Output power: 0.7W
Loudspeaker: 80mm, circular
Frequency range: 100Hz-8kHz
Tone controls: No
Recording level: Automatic
Microphone: Built-in
Monitor on record: No
Inputs: None
Outputs: None
Pause: Yes
Cue and review: Yes
Tape counter: No
Power supply: Mains/battery
Weight: 0.88kg
Dimensions: 139 x 48 x 251mm
Cost: nia
Accessories: None

Make: **Neal**
Model: 102
Output power: Line only
Loudspeaker: No
Frequency range: 35Hz-15kHz
Tone controls: No
Recording level: Manual (metered)
Microphone: External
Monitor on record: nia
Inputs: Mic; high level line; low level line
Outputs: Line; headphones
Pause: Yes
Cue and review: nia
Tape counter: 3 digit
Power supply: nia
Weight: 6.4kg
Dimensions: 315 x 214 x 113mm
Cost: nia
Accessories: None
Other models: 102V – as above with variable tape bias. 103 – as 102V with input mixing facilities
Notes: Can be converted for high speed copying

Make: **Neal**
Model: 140
Output power: Line only
Loudspeaker: No
Frequency range: 35Hz-15kHz
Tone controls: No
Recording level: Manual (metered)
Microphone: External
Monitor on record: nia
Inputs: Mic; line (4 tracks simultaneously)
Outputs: Line; headphones
Pause: Yes
Cue and review: nia
Tape counter: 3 digit
Power supply: nia
Weight: 7.3kg
Dimensions: 460 x 241 x 191mm
Cost: nia
Accessories: None
Notes: Dolby noise reduction; full tape width (4 track) record/playback head; can be converted for high speed copying

Make: **Neal**
Model: 312 Stereo
Output power: Line only
Loudspeaker: No
Frequency range: 35Hz-15kHz
Tone controls: No
Recording level: Manual (metered)
Microphone: External
Monitor on record: nia
Inputs: Mic; high level line; low level line
Outputs: Line; headphones
Pause: No
Cue and review: No
Tape counter: 3 digit
Power supply: 105-125V or 220-250V
Weight: 8kg
Dimensions: 446 x 226 x 155mm
Cost: nia
Accessories: None
Notes: Dolby noise reduction; stereo or mono recording

Make: **Philips**
Model: N 2233
Output power: 0.8W
Loudspeaker: Integral
Frequency range: nia
Tone controls: No
Recording level: Automatic
Microphone: Built-in
Monitor on record: No
Inputs: Mic; line
Outputs: nia
Pause: No
Cue and review: Yes
Tape counter: No
Power supply: Mains/battery
Weight: nia
Dimensions: 177 x 76 x 270mm
Cost: (a)
Accessories: None
Other models: N 2234 – as above with ext speaker socket; tone control; 3 digit counter; pause control (a)

Make: **Philips**
Model: N 2235
Output power: 1.5W
Loudspeaker: Integral
Frequency range: nia
Tone controls: Treble cut
Recording level: Automatic
Microphone: Built-in
Monitor on record: No
Inputs: Mic; line
Outputs: Ext speaker
Pause: Yes
Cue and review: Yes
Tape counter: 3 digit
Power supply: Mains/battery
Weight: nia
Dimensions: 200 x 73 x 290mm
Cost: (a)
Accessories: None

Make: **Rank Aldis**
Model: 148B
Output power: 1W
Loudspeaker: 100mm, elliptical
Frequency range: 60Hz-10kHz
Tone controls: Treble cut
Recording level: Automatic
Microphone: Built-in/external
Monitor on record: No
Inputs: Mic; line
Outputs: Headphones
Pause: Yes
Cue and review: Yes
Tape counter: 3 digit
Power supply: 240V AC or 4 x 1.5V DC
Weight: 2kg
Dimensions: 217 x 224 x 76mm
Cost: (a)
Accessories: Microphone; headphones

Make: **Rank Aldis**
Model: 244
Output power: 750mW
Loudspeaker: 75mm, round
Frequency range: 100Hz-8kHz
Tone controls: Switched
Recording level: Automatic
Microphone: Built-in
Monitor on record: No
Inputs: Mic; line
Outputs: Headphones
Pause: No
Cue and review: Yes
Tape counter: 3 digit
Power supply: 240V AC or 4 x 1.5V DC
Weight: 1.8kg

Dimensions: 229 x 216 x 83mm
Cost: (a)
Accessories: Microphone; headphones

Make: **Revox**
Model: D-88
Output power: nia
Loudspeaker: No
Frequency range: 60Hz-12kHz
Tone controls: No
Recording level: Automatic
Microphone: External
Monitor on record: No
Inputs: Mic; remote
Outputs: Headphones; speaker
Pause: Yes
Cue and review: nia
Tape counter: 3 digit
Power supply: 100-240V
Weight: 8.5kg
Dimensions: 210 x 143 x 410mm
Cost: (e)
Accessories: Remote control; speaker
Notes: Quadruple speed copying possible

Make: **Tandberg**
Model: Audio Tutor 771
Output power: 3W
Loudspeaker: 165mm, circular
Frequency range: nia
Tone controls: Treble cut
Recording level: Automatic/manual (metered)
Microphone: Built-in
Monitor on record: Yes
Inputs: Mic; line; remote
Outputs: Ext speaker; line
Pause: Yes
Cue and review: Yes
Tape counter: 3 digit
Power supply: 240V AC or 9V DC
Weight: 3.75kg
Dimensions: 222 x 340 x 95mm
Cost: (a)
Accessories: Microphone

Make: **Tandberg**
Model: TCR 222
Output power: 12W
Loudspeaker: Integral
Frequency range: 40Hz-14kHz
Tone controls: Bass and treble
Recording level: Manual (metered)
Microphone: External
Monitor on record: Yes
Inputs: Mic; radio; phono
Outputs: Radio; ext speaker
Pause: Yes
Cue and review: No
Tape counter: 3 digit
Power supply: 230V
Weight: 6kg
Dimensions: Height – 11mm, width – 435mm
Cost: (b)
Accessories: Microphone
Other models: TCR 222 SL, language practice recorder with audio-active headset (b)

Make: **Technics (National Panasonic)**
Model: RS 263 AUS
Output power: Line only
Loudspeaker: No
Frequency range: 30Hz-15kHz
Tone controls: No
Recording level: Manual (metered)
Microphone: External
Monitor on record: Yes
Inputs: Mic; line
Outputs: Line; headphones
Pause: Yes
Cue and review: No
Tape counter: 3 digit
Power supply: 90-250V
Weight: 4.8kg
Dimensions: 375 x 120 x 242mm
Cost: nia
Accessories: None
Notes: Stereo model with Dolby noise reduction

Make: **Technics (National Panasonic)**
Model: RS 640 USD
Output power: Line only
Loudspeaker: No
Frequency range: 20Hz-16kHz
Tone controls: No
Recording level: Manual line and mic
Microphone: External
Monitor on record: Yes
Inputs: Mic; line
Outputs: Line; headphones
Pause: Yes
Cue and review: No
Tape counter: 3 digit
Power supply: 240V
Weight: 5.8kg
Dimensions: 432 x 130 x 301mm
Cost: nia
Accessories: Timer switch
Notes: Stereo model with Dolby noise reduction

Make: **Telex (Avcom Systems Ltd)**
Model: C-110
Output power: 10W
Loudspeaker: Integral
Frequency range: 40Hz-10kHz
Tone controls: Yes
Recording level: Automatic/manual
Microphone: External (supplied)
Monitor on record: No
Inputs: Mic; line
Outputs: nia
Pause: Yes
Cue and review: Yes
Tape counter: Yes
Power supply: 220V
Weight: 6.8kg
Dimensions: 133 x 254 x 374mm
Cost: (b)
Accessories: Foot switch; headphones; headset

Make: **Wollensak (3M)**
Model: 2520 AV
Output power: 9W
Loudspeaker: Integral
Frequency range: 50Hz-10kHz
Tone controls: Treble cut
Recording level: Automatic
Microphone: External (supplied)
Monitor on record: No
Inputs: Mic; line
Outputs: Line; ext speaker
Pause: Yes
Cue and review: No
Tape counter: 3 digit
Power supply: 120-240V
Weight: 7.26kg
Dimensions: 314 x 365 x 165mm
Cost: (d)
Accessories: None

LANGUAGE LABORATORIES

There are two basic systems used in the design of language laboratories. The audio active comparative (AAC) system allows the student to hear the master recording, record his response, then compare it to the master track. In the audio active (AA) system the student can hear both master track and his own response in his headset. Since there are many variations and improvements continually being made in the design of language laboratories, and since the installation is generally an expensive one, it is advisable for any potential purchaser to seek advice both from manufacturers and existing users.

Make: **Clarke and Smith**
Model: 108 AA
Tape:
Tracks:
Recording level:
Cue and review:
Tape counter: } No integral tape recording system
Programme channels: 3 plus teacher microphone
Programme sources: 4 sockets to connect to any programme source and microphone
Max no of student units: 40
Student grouping: 8 groups of 5
Remote control facilities: All call; group call; teacher call; intercom
Power supply: 200-250V
Dimensions: Control console; 876 x 1219 x 635mm; control unit; 444 x 305 x 190mm; student amplifier; 279 x 760mm (circular)
Cost: (k)
Accessories: None

Make: **Clarke and Smith**
Model: Control console panel – MT6, individual study machine – MT6M
Tape: Cassette
Tracks: Lower ½ – master; upper ½ – student
Recording level: Manual
Cue and review: No
Tape counter: 3 digit
Programme channels: 3
Programme sources: Mic; master recorder; external recorder
Max no of student units: 30
Student grouping: 1 group operation
Remote control facilities: All call; monitor; intercom; recording level; test recording
Power supply: 240V
Dimensions: Control console – 330 x 110 x 307mm; individual machine – 300 x 211 x 105mm
Cost: nia
Accessories: None
Notes: Operating modes AAC and listening via loudspeaker. Individual machines can be disconnected for private study

Make: **ESL**
Model: Flexilab 2 (Senior or Junior)
Tape: Cassette
Tracks: Lower ½ – master; upper ½ – student
Recording level: Automatic
Cue and review: Yes
Tape counter: 3 digit
Programme channels: Senior – 2; junior – 1
Programme sources: Any audio source
Max no of student units: Senior – 40; junior 10 (or up to 40 by linking consoles)
Student grouping: Senior – up to 5 groups of 8
Remote control facilities: Senior – all call; group call; intercom; group conference; programme selection; monitoring; junior – all call; monitoring and communication to any selected student; programme transfer
Power supply: nia
Dimensions: Senior console – (16 students) – 500 x 300 x 120mm; (32 students) – 830 x 310 x 120mm; (40 students) – 999 x 310 x 120mm; junior console – 300 x 260 x 100mm; flexible 2 recorder – 330 x 285 x 90mm. Power supply (senior) – 375 x 230 x 460mm; (junior) – 180 x 210 x 200mm
Cost: nia
Accessories: None
Notes: Furniture and fittings can be supplied

Make: **Friedelab**
Model: 11-41CC
Tape: Cassette or open reel
Tracks: Half-track
Recording level: VU meter
Cue and review: nia
Tape counter: Yes
Programme channels: 3
Programme sources: Open reel; cassette; radio; disc; microphone; monitored student
Max no of student units: 32
Student grouping: nia
Remote control facilities: All call; teacher call; monitor; intercom; group conference
Power supply: 240V
Dimensions: nia
Cost: nia
Accessories: Tape storage unit; loudspeaker; spares kit

Make: **Revox**
Model: C88 Monoprogram
Tape: Cassette (Revox C88)
Tracks: Lower ½ – master; upper ½ – student
Recording level: Automatic
Cue and review: Automatic sentence repetition
Tape counter: 3 digit
Programme channels: 1
Programme sources: Up to 4, cassette, open reel, mic and auxiliary
Max no of student units: 36
Student grouping: 1 group operation
Remote control facilities: All call; teacher call; intercom; monitoring; conference (up to 4 students); test recording
Power supply: 110-250V
Dimensions: Console – 1248 x 780 x 600mm. Student's desk – 710 x 780 x 600mm
Cost: nia
Accessories: None
Notes: Operating modes AAC:AA; listening via loudspeaker

Make: **Revox**
Model: 884
Tape: Cassette
Tracks: Lower ½ – master; upper ½ – student
Recording level: Automatic
Cue and review: Automatic sentence repetition
Tape counter: 3 digit (electronic)
Programme channels: 4 (can be increased to 10)
Programme sources: Cassette; open reel; mic; auxiliary. (6

extra channels can be provided)
Max no of student units: 75
Student grouping: 2 groups any number (2 more groups can be added)
Remote control facilities: All call; group call; teacher call; intercom; monitoring; fast forward; fast rewind; play; test recording
Power supply: 100-240V
Dimensions: Console – 1330 x 870 x 560mm; student's desk – 720 x 780 x 560mm
Cost: nia
Accessories: None
Notes: Expansion to dual function possible

Make: **Smiths Electrical Engineers**
Model: Sebitron Digital Lab System
Tape: Cassette
Tracks: Lower ¼ – master; next ¼ – student
Recording level: Pre-set or manual
Cue and review: Yes
Tape counter: 3 digit
Programme channels: 2
Programme sources: Up to 2 tape and 1 mic
Max no of student units: 40
Student grouping: 2 groups of any number
Remote control facilities: All call; group call; intercom; monitor; test recording
Power supply: 200-250V
Dimensions: Portable console; 500 x 500 x 215mm; fixed console; 1400 x 740 x 760mm
Cost: nia
Accessories: None
Notes: Available with either portable or fixed console

Make: **Stillitron**
Model: MT6 Minilab
Tape: Cassette
Tracks: Lower ½ – master; upper ½ – student
Recording level: Automatic
Cue and review: Semi-automatic repeat facility
Tape counter: 3 digit
Programme channels: 1
Programme sources: Any audio source
Max no of student units: 30
Student grouping: 1
Remote control facilities: All call; intercom; monitoring
Power supply: 220/240V
Dimensions: Console – 298 x 330 x 101mm; student's unit – 300 x 211 x 105mm
Cost: nia
Accessories: None
Notes: AA and AAC facility

Make: **Stillitron**
Model: SP3 console/CP760 recorder
Tape: Cassette
Tracks: Lower ½ – master; upper ½ – student
Recording level: Automatic (manual adjust at console)
Cue and review: Rewind and automatic play
Tape counter: 3 digit
Programme channels: 4 plus teacher's plus live programme
Programme sources: Cassette; open reel; turntable; timer
Max no of student units: 48
Student grouping: Groups of 8 up to 6 groups
Remote control facilities: Full remote facilities
Power supply: 110V, 220/240V
Dimensions: nia
Cost: nia
Accessories: None
Notes: Modular plug-in construction

Make: **Stillitron**
Model: SP25 console and student's recorder
Tape: Cassette
Tracks: Lower ½ – master; upper ½ – student
Recording level: Automatic
Cue and review: No
Tape counter: 3 digit
Programme channels: 3
Programme sources: Cassette; open reel; turntable, tuner
Max no of student units: 30
Student grouping: 3
Remote control facilities: All call; group call; conference; intercom; fast forward; eject and lock; record; play
Power supply: 110V, 220/240V
Dimensions: Console – 550 x 340 x 120mm; recorder – 230 x 290 x 90mm
Cost: nia
Accessories: None
Notes: AA or AAC programme facility

Make: **Sony**
Model: LLC 1000 Translab with ER1000 Recorder
Tape: Cassette
Tracks: Lower ¼ – master; next ¼ – student, or lower ½ – master; upper ½ – student
Recording level: Automatic
Cue and review: Yes
Tape counter: 3 digit
Programme channels: 1
Programme sources: Any audio source
Max no of student units: 10 (up to 30 by connecting 3 x LLC 1000)
Student grouping: 1
Remote control facilities: All call; teacher call; intercom; conference; monitoring; test recording
Power supply: 220V
Dimensions: LLC 1000 – 388 x 118 x 142mm; ER 1000 – 240 x 88 x 180mm
Cost: nia
Accessories: None
Notes: Student booth amplifier EASH compatible with LLC 1000 and ER 1000

Make: **Sony**
Model: LLC-6 Learning Lab System
Tape: Cassette
Tracks: Lower ¼ – master; next ¼ – student, or lower ½ – master; upper ½ – student
Recording level: Automatic
Cue and review: No
Tape counter: 3 digit
Programme channels: 2 (for each LU-6020 unit)
Programme sources: Any audio source
Max no of student units: 10 per LU-6020 unit (up to 7 with 7 x LU-6020)
Student grouping: 1 per RM-1010 unit (fixed once arranged)
Remote control facilities: With remote units RM-1020 and RM-1010 full remote control is possible
Power supply: 240V
Dimensions: nia
Cost: nia
Accessories: None
Notes: The system comprises 2 console units (LU-6010 and LU-6020), student booth amplifier (EASH), student cassette recorder (ER-740, ER-780 or ER-840), master remote control (RM-1020), group remote control (RM-1010)

Make:	**Tandberg**
Model:	IS9
Tape:	Cassette
Tracks:	Lower ½ – master; upper ½ – student
Recording level:	Automatic
Cue and review:	Gap seeking and continuous
Tape counter:	3 digit
Programme channels:	2
Programme sources:	Cassette; open reel; or any external audio source
Max no of student units:	64
Student grouping:	2
Remote control facilities:	All call; teacher call; group call; intercom; monitor; conference; programme transfer; high speed transfer; auto transfer; test recording
Power supply:	240V
Dimensions:	nia
Cost:	nia
Accessories:	None

LOUDSPEAKERS

The main use of a separate loudspeaker is to improve the audio output of equipment such as radios or cassette recorders when these are being used with a large group of students. Most portable audio amplifiers tend to have built-in speakers which are too small to do justice to the power available. As a result, a noticeable improvement can be achieved simply by connecting an external speaker. Of course, the external speaker does not increase the power output of the equipment, it merely handles the available power without distortion. The power handling capability of the extension speaker should be slightly in excess of the maximum power output of the amplifier while the frequency range which it can handle will determine whether it is suitable for speech or music. Generally, a speaker intended for speech only will need a range of about 150Hz to 8kHz, while for good quality music reproduction a range of from 30Hz to 16kHz would be tolerable but an extended range would be better.

Make:	**Audix**
Model:	C2/Hz
Max power handling:	2W
Impedance:	100V line
Frequency range:	80Hz-14kHz
Units:	1
Mounting:	Wall or free standing
Weight:	1.15kg
Dimensions:	135 x 225 x 70mm
Cost:	(a)
Other models:	C2/H/VC – as above with volume control
Notes:	Transformer tapped at 1W and 0.5W

Make:	**Audix**
Model:	C4/Hz
Max power handling:	4W
Impedance:	100V line
Frequency range:	60Hz-11kHz
Units:	1
Mounting:	Wall or free standing
Weight:	2kg
Dimensions:	190 x 315 x 85mm
Cost:	(a)
Other models:	C4/Hz/VC – as above with volume control
Notes:	Transformer tapped at 2W and 1W

Make:	**Audix**
Model:	C8/Hz
Max power handling:	8W
Impedance:	100V line
Frequency range:	50Hz-10kHz
Units:	1
Mounting:	Wall and free standing
Weight:	4kg
Dimensions:	225 x 405 x 115mm
Cost:	(a)
Other models:	C8/Hz/VC – as above with volume control
Notes:	Transformer tapped at 4W and 2W

Make:	**Audix**
Model:	CL 120
Max power handling:	12W
Impedance:	8 ohms or 100V line (on order)
Frequency range:	60H-15H
Units:	nia
Mounting:	Wall
Weight:	3.8kg
Dimensions:	790 x 165 x 76mm
Cost:	(a)
Other models:	CL 200 – as above but larger and handling 20W (b). CL 320 – as above but larger and handling 32W (b)

Make:	**Clarke and Smith**
Model:	H10
Max power handling:	6W
Impedance:	15 ohms
Frequency range:	nia
Units:	1 – 254mm diameter
Mounting:	Wall or portable
Weight:	4.5kg
Dimensions:	380 x 380 x 222mm
Cost:	(a)
Other models:	H10 plus T – as above with line transformer fitted (a). H10 plus VC – as above with volume control fitted (a)

Make:	**Clarke and Smith**
Model:	L488
Max power handling:	6W
Impedance:	8 ohms and 70/100V line
Frequency range:	nia
Units:	7 – 127mm diameter
Mounting:	Wall
Weight:	9.9kg
Dimensions:	952 x 203 x 165mm
Cost:	(a)
Other models:	L437 – – as above but 7W. L437/12 – as above but 12W

Make:	**Coomber**
Model:	F10
Max power handling:	10W
Impedance:	8, 15 ohms or 70/100V line (to order)
Frequency range:	nia
Units:	1 – 254mm diameter
Mounting:	Free standing, portable
Weight:	2.8kg
Dimensions:	305 x 356 x 115mm
Cost:	(a)
Other models:	W10 – as above but wall mounting (a)
Notes:	Optional on/off switch or volume control

Make: **Coomber**
Model: C014/2HF
Max power handling: 40W
Impedance: 8, 15 ohms or 70/110V line (to order)
Frequency range: nia
Units: 4 – 254mm diameter plus 2 high frequency units
Mounting: Wall
Weight: 11.36kg
Dimensions: 1175 x 320 x 133mm
Cost: (b)
Notes: Optional on/off switch or volume control

Make: **Coomber**
Model: F20
Max power handling: 20W
Impedance: 8, 15 ohms or 70/100V line (to order)
Frequency range: nia
Units: 2 – 254mm diameter
Mounting: Free standing, portable
Weight: 5.9kg
Dimensions: 665 x 387 x 115mm, without stand
Cost: (a)
Other models: W20 – as above but wall mounting (a)
Notes: Optional on/off switch or volume control

Make: **Eagle**
Model: CS 105
Max power handling: 5W
Impedance: 100V line
Frequency range: 200Hz-10kHz
Units: 5 – 90mm diameter
Mounting: Wall or free standing
Weight: nia
Dimensions: 585 x 140 x 75mm
Cost: (a)
Other models: A series of 5 speakers handling up to 25W (a) to (b). CL series, 3 speakers, handling from 5W-30W (a) to (b). WS series, 7 speakers, handling from 2W-8W (a)

Make: **Goodsell**
Model: P10
Max power handling: 10W
Impedance: 8-15 ohms
Frequency range: nia
Units: 1 – 254mm diameter
Mounting: Free standing, portable
Weight: 5.4kg
Dimensions: 394 x 381 x 222mm
Cost: (a)
Other models: HP10 – as above with 70/100V line transformer (a)

Make: **Goodsell**
Model: LS8
Max power handling: 15W
Impedance: 8 ohms
Frequency range: nia
Units: 1 – 102mm diameter
Mounting: Free standing, portable
Weight: 6kg
Dimensions: 457 x 290 x 280mm
Cost: (a)

Make: **Goodsell**
Model: P12
Max power handling: 25W
Impedance: 8 ohms
Frequency range: nia
Units: 1 – 305mm diameter plus 1 high frequency unit
Mounting: Free standing, portable
Weight: 10.5kg
Dimensions: 533 x 368 x 228mm
Cost: (a)

Make: **Rediffusion**
Model: RWM 1
Max power handling: 1W
Impedance: 100V line
Frequency range: nia
Units: 1
Mounting: Wall
Weight: 0.9kg
Dimensions: 127 x 203 x 76mm
Cost: (a)
Other models: RWM 2, RWM 5 and RWM 8; each speaker in the series becoming larger and heavier. Power handling 2W, 5W and 8W (a)

Make: **Rediffusion**
Model: RLC-5
Max power handling: 5W
Impedance: 100V line
Frequency range: nia
Units: 2 – 125mm diameter
Mounting: Wall
Weight: 3.4kg
Dimensions: 533 x 165 x 100mm
Cost: (a)
Other models: RLC-10, RLC-20; each speaker in the series becoming larger and heavier with more units. Power handling 10W and 20W (a)

Make: **Tandberg**
Model: TL 25-20
Max power handling: 65W
Impedance: 4 ohm
Frequency range: 48Hz-20kHz
Units: 1 treble, 1 bass
Mounting: Free standing
Weight: nia
Dimensions: 419 x 266 x 222mm
Cost: (a)

Make: **Tandberg**
Model: TL 35-20
Max power handling: 75W
Impedance: 4-8 ohms
Frequency range: 38Hz-20kHz
Units: 1 bass, 1 mid-range, 1 treble
Mounting: Free standing
Weight: nia
Dimensions: 590 x 355 x 260mm
Cost: (a)

Make: **TOA**
Model: T-101
Max power handling: 15W
Impedance: 1000 ohms
Frequency range: 160H-10kH
Units: 2 – 178 x 127mm elliptical
Mounting: Wall
Weight: 5.9kg
Dimensions: 167 x 466 x 141mm
Cost: (a)
Other models: A complete range of 10 column speakers with similar characteristics handling up to 60W (a) to (c)

Make: **TOA**
Model: BS 364

Max power handling:	4.5W
Impedance:	3300 ohms
Frequency range:	100Hz-8kHz
Units:	1 – 160mm diameter
Mounting:	Wall
Weight:	1.45kg
Dimensions:	260 x 260 x 130mm
Cost:	(a)
Other models:	A complete range of 11 box speakers with similar characteristics handling up to 9W (a)

OPEN REEL TAPE RECORDERS

Open reel tape recorders use ¼ inch magnetic tape on spools. The tape has to be threaded through the recorder and past the erase and recording heads. The tape can be transported at speeds which vary from 38cm/s to 2.4cm/s. The two most common speeds are 19cm/s and 9.5cm/s. Generally the higher tape speeds result in improved high frequency response in the recording with a noticeable reduction in background noise. Most of the modern recorders in this class are high quality ones since the advent of the cassette recorder has filled the space for machines in the lower price brackets.

Most of the machines listed have integral amplifiers and loudspeakers but a few have been included without these. The number of tracks available for recording vary. Mono recorders can be supplied with either two or four track heads while for stereo recording it is only possible to have a two track format. While some 4 track mono recorders can play back stereo recordings, in mono, a two track mono machine cannot replay a stereo recording.

Make:	**Akai**
Model:	1722 II
Max spool size:	180mm
Speeds:	19, 9.5cm/s
Tracks:	4, 2 channel stereo
Output power:	5 plus 5W
Loudspeaker:	Integral
Frequency range:	30Hz-21kHz at 19cm/s
Tone controls:	No
Recording level:	Manual (metered)
Microphone:	External
Monitor on record:	Yes
Inputs:	Mic; line
Outputs:	Line; headphones
Pause:	Yes
Tape counter:	4 digit
Power supply:	Mains
Weight:	13.2kg
Dimensions:	358 x 360 x 248mm
Cost:	(c)
Accessories:	None

Make:	**Akai**
Model:	GX 4000 D
Max spool size:	180mm
Speeds:	19, 9.5cm/s
Tracks:	4, 2 channel stereo
Output power:	Deck – no power amplifiers
Loudspeaker:	None
Frequency range:	30Hz-24kHz at 19cm/s
Tone controls:	No
Recording level:	Manual (metered)
Microphone:	External
Monitor on record:	Yes
Inputs:	Mic; line
Outputs:	Line; headphones
Pause:	Yes
Tape counter:	4 digit
Power supply:	Mains
Weight:	13.2kg
Dimensions:	440 x 315 x 230mm
Cost:	(c)
Accessories:	None
Other models:	GX 4000 DB – as above with Dolby noise reduction (c)
Notes:	Sound-on-sound facility

Make:	**Akai**
Model.	GX 620 D
Max spool size:	266mm
Speeds:	19, 9.5cm/s
Tracks:	4, 2 channel stereo
Output power:	Deck – no power amplifiers
Loudspeaker:	None
Frequency range:	30Hz-26kHz at 19cm/s
Tone controls:	No
Recording level:	Manual (metered)
Microphone:	External
Monitor on record:	Yes
Inputs:	Mic; line
Outputs:	Line; headphones
Pause:	Yes
Tape counter:	4 digit
Power supply:	Mains
Weight:	17.6kg
Dimensions:	440 x 315 x 230mm
Cost:	(e)
Accessories:	None
Notes:	Mic/line mixing

Make:	**Akai**
Model:	GX 635 D
Max spool size:	266mm
Speeds:	19, 9.5cm/s
Tracks:	4, 2 channel stereo
Output power:	Deck – no power amplifiers
Loudspeaker:	None
Frequency range:	30Hz-27kHz at 19cm/s
Tone controls:	No
Recording level:	Manual (metered)
Microphone:	External
Monitor on record:	Yes
Inputs:	Mic; line
Outputs:	Line; headphones
Pause:	Yes
Tape counter:	Real time
Power supply:	Mains
Weight:	21kg
Dimensions:	440 x 483 x 256mm
Cost:	(g)
Accessories:	None
Notes:	Auto reverse on record/replay; 6 heads

Make:	**Ferrograph (Wilmot Breeden Electronics)**
Model:	Logic 7 series
Max spool size:	270mm
Speeds:	38, 19, 9.5cm/s
Tracks:	2 or 4
Output power:	2 x 10W
Loudspeaker:	2 x 180 x 100mm
Frequency range:	30Hz-20kHz at 38cm/s
Tone controls:	Bass and treble

Recording level: Manual (metered)
Microphone: External
Monitor on record: Yes
Inputs: Mic; line
Outputs: Line; headphones; ext speakers
Pause: Yes
Tape counter: 4 digit
Power supply: 220-250V
Weight: 26kg
Dimensions: 515 x 445 x 255mm
Cost: nia
Accessories: None
Other models: 7602H, ½ track stereo recorder only. 7604H, ¼ track stereo recorder only. 7622H, ½ track with power amplifier and loudspeaker. 7624H, ¼ track with power amplifier and loudspeaker. All the above available with or without Dolby noise reduction

Make: **Philips**
Model: N 4504
Max spool size: 180mm
Speeds: 19, 9.5, 4.76cm/s
Tracks: 4
Output power: Deck – no power amplifier
Loudspeaker: None
Frequency range: 35Hz-25kHz at 19cm/s
Tone controls: No
Recording level: Manual (metered)
Microphone: External
Monitor on record: Yes
Inputs: Mic; line
Outputs: Line; monitor
Pause: Yes
Tape counter: 4 digit
Power supply: 110, 127, 220, 240V
Weight: 8.25kg
Dimensions: 415 x 433 x 195mm
Cost: nia
Accessories: None
Notes: Stereo

Make: **Philips**
Model: N 4506
Max spool size: 180mm
Speeds: 19, 9.5, 4.76cm/s
Tracks: 4
Output power: Deck – no power amplifier
Loudspeaker: None
Frequency range: 35Hz-25kHz at 19cm/s
Tone controls: Bass and treble
Recording level: Manual (metered)
Microphone: External
Monitor on record: Yes
Inputs: Mic; line; tuner; phono; auxiliary (all separately controlled for mixing)
Outputs: Line; headphones
Pause: Yes
Tape counter: 4 digit
Power supply: 110, 127, 220, 240V
Weight: 14kg
Dimensions: 557 x 437 x 210mm
Cost: nia
Accessories: None
Notes: Stereo

Make: **Pioneer**
Model: RT-707
Max spool size: 180mm
Speeds: 19, 9.5cm/s
Tracks: 4, 2 channel stereo
Output power: Deck – no power amplifiers
Loudspeaker: None
Frequency range: 20Hz-20kHz at 19cm/s
Tone controls: No
Recording level: Manual (metered)
Microphone: External
Monitor on record: nia
Inputs: nia
Outputs: nia
Pause: nia
Tape counter: 4 digit
Power supply: Mains
Weight: 20kg
Dimensions: 480 x 230 x 356mm
Cost: (e)
Accessories: None

Make: **Revox**
Model: B 77 series
Max spool size: 268mm
Speeds: 38 and 19cm/s or 19 and 9.5cm/s
Tracks: 2 or 4
Output power: Deck – no power amplifiers
Loudspeaker: None
Frequency range: 30Hz-22kHz at 38cm/s
Tone controls: No
Recording level: Manual (metered)
Microphone: External
Monitor on record: Yes
Inputs: Mic (low/high); line; radio
Outputs: Line; radio; headphones; remote control
Pause: Yes
Tape counter: 4 digit
Power supply: 100, 120, 140, 200, 220, 240V
Weight: 17kg
Dimensions: 452 x 414 x 207mm
Cost: (g)-(h)
Accessories: None
Other models: 42 different versions depending on tape speed, reel size, and case trim

Make: **Sony**
Model: TC 399
Max spool size: 180mm
Speeds: 19, 9.5, 4.76cm/s
Tracks: 4, 2 channel
Output power: Deck – no power amplifier
Loudspeaker: None
Frequency range: 30Hz-25kHz at 19cm/s
Tone controls: No
Recording level: Manual (metered)
Microphone: External
Monitor on record: Yes
Inputs: Mic; line
Outputs: Line; headphones
Pause: Yes
Tape counter: 4 digit
Power supply: 110, 120, 220, 240V
Weight: 12.9kg
Dimensions: 415 x 435 x 190mm
Cost: (c)
Accessories: None
Notes: Stereo

Make: **Sony**
Model: TC 765
Max spool size: 268mm
Speeds: 19, 9.5cm/s
Tracks: 4, 2 channel
Output power: Deck – no power amplifier
Loudspeaker: None
Frequency range: 30Hz-25kHz at 19cm/s
Tone controls: None
Recording level: Manual (metered)
Microphone: External
Monitor on record: Yes

Inputs:	Mic; line
Outputs:	Line; headphones
Pause:	Yes
Tape counter:	4 digit
Power supply:	110, 120, 220, 240V
Weight:	27kg
Dimensions:	445 x 525 x 235mm
Cost:	(f)
Accessories:	None
Other models:	TC766-2 – as above with 2 track record/playback facility (g)
Notes:	Stereo

Make:	**Sony**
Model:	TC 880-2
Max spool size:	268mm
Speeds:	38, 19cm/s
Tracks:	2, 2 channel or 4, 2 channel
Output power:	Deck – no power amplifier
Loudspeaker:	None
Frequency range:	20Hz-40kHz at 38cm/s
Tone controls:	No
Recording level:	Manual (metered)
Microphone:	External
Monitor on record:	Yes
Inputs:	Mic; line
Outputs:	Line; headphones
Pause:	Yes
Tape counter:	4 digit
Power supply:	110, 120, 220, 240V
Weight:	36.5kg
Dimensions:	465 x 515 x 265mm
Cost:	(k)
Accessories:	None

Make:	**Tandberg**
Model:	1521
Max spool size:	180mm
Speeds:	19, 9.5, 4.76cm/s
Tracks:	2
Output power:	10W
Loudspeaker:	178 x 100mm, elliptical
Frequency range:	40Hz-19kHz at 19cm/s
Tone controls:	Bass and treble
Recording level:	Manual (metered)
Microphone:	External
Monitor on record:	Yes
Inputs:	Mic; line
Outputs:	Line; ext speaker
Pause:	Yes
Tape counter:	4 digit
Power supply:	115, 230, 240V
Weight:	10.3kg
Dimensions:	400 x 171 x 333mm
Cost:	(c)
Accessories:	Remote control; foot switch
Other models:	1521F – as above with remote control facilities (c). TFC3 – as 1521F with foot control (c). 1541 – as 1521 but 4 track (c)

Make:	**Technics (National Panasonic)**
Model:	RS-1500 US
Max spool size:	266mm
Speeds:	38cm/s
Tracks:	4
Output power:	Deck – no power amplifiers
Loudspeaker:	None
Frequency range:	30Hz-30kHz at 38cm/s
Tone controls:	No
Recording level:	Manual (metered)
Microphone:	External
Monitor on record:	Yes
Inputs:	Mic; line
Outputs:	Line; headphones
Pause:	Yes
Tape counter:	Elapsed time
Power supply:	110-240V
Weight:	25kg
Dimensions:	456 x 446 x 258mm
Cost:	nia
Accessories:	None
Other models:	RS-1700 – as above with automatic reverse

PUBLIC ADDRESS AMPLIFIERS

These units are intended to amplify speech or music to a level where it would be useful for large audiences. They do not have built-in loudspeakers, since their power output is generally high and varies from around 20 watts to over 500 watts. Most have output sockets for the standard loudspeaker impedance of 4 ohms to 16 ohms, while many have additional output terminals which permit the connection of a series of loudspeakers to a 70 volt or 100 volt line. This later type of connection is necessary when a large number of speakers spread over a large area are required.

Make:	**Amcron (MacInnes Labs Ltd)**
Model:	D-75
Output power:	60W per channel
Frequency range:	20Hz-20kHz
Tone controls:	No
Inputs:	Line
Outputs:	Speaker; headphones
Power supply:	100-240V
Weight:	4.5kg
Dimensions:	483 x 222 x 34mm
Cost:	(c)
Accessories:	None
Notes:	Stereo/monaural

Make:	**Amcron (MacInnes Labs Ltd)**
Model:	D 150 A
Output power:	190W per channel
Frequency range:	1Hz-20kHz
Tone controls:	No
Inputs:	Line
Outputs:	Speakers
Power supply:	120-240V
Weight:	11.4kg
Dimensions:	432 x 133 x 222mm
Cost:	(d)
Accessories:	None
Notes:	Stereo/monaural

Make:	**Amcron (MacInnes Labs Ltd)**
Model:	DC 300 A
Output power:	500W per channel
Frequency range:	1Hz-20kHz
Tone controls:	No
Inputs:	Line
Outputs:	Speakers
Power supply:	120-240V
Weight:	22kg
Dimensions:	483 x 178 x 248mm
Cost:	(i)

Accessories: None
Notes: Stereo/monaural

Make: **Eagle**
Model: TPA 35
Output power: 10W
Frequency range: nia
Tone controls: Treble
Inputs: 2 mic; 2 line
Outputs: 4, 8 and 16 ohms; 25 and 100V line
Power supply: 220-240V AC or 12V DC
Weight: nia
Dimensions: 95 x 255 x 230mm
Cost: (a)
Accessories: None
Other models: TPA 40N – as above with 20W output (a).
TPA 45 – as above with 30W output (a).
TPA 60 – as above with 50W output (b)

Make: **Goodsell**
Model: PA 450
Output power: 50W
Frequency range: 50Hz-20kHz
Tone controls: Bass and treble
Inputs: 2 mic, plus 2 line
Outputs: 100V line
Power supply: 240V
Weight: 8.6kg
Dimensions: 373 x 240 x 90mm
Cost: (b)
Accessories: None

Make: **TOA**
Model: TA 403E
Output power: 30W
Frequency range: 50Hz-15kHz
Tone controls: Bass and treble
Inputs: 3 mic; 3 line; 1 magnetic
Outputs: 4 and 16 ohms; 50, 70 and 100V line
Power supply: 240V AC or 24V DC
Weight: 8.6kg
Dimensions: 420 x 272 x 152mm
Cost: (b)
Accessories: None
Other models: TA 406E – as above with 60W output (c).
TA 412E – as above with 120W output (d)

Make: **TOA**
Model: TA 301
Output power: 15W
Frequency range: 100Hz-15kHz
Tone controls: Speech/music switch
Inputs: 2 mic; 2 line
Outputs: 4 ohms, 50, 70 and 100V line
Power supply: 240V AC or 24V DC
Weight: 3.6kg
Dimensions: 342 x 128 x 198mm
Cost: (b)
Accessories: None
Other models: TA 303 – as above with 30W output (b)

Make: **Vortexion (Clarke and Smith)**
Model: VTN CP50W
Output power: 50W
Frequency range: 30Hz-14kHz
Tone controls: Bass and treble
Inputs: 4
Outputs: 100V line; 8 and 15 ohms
Power supply: Mains and/or 12V DC
Weight: 16kg
Dimensions: 356 x 292 x 115mm
Cost: (c)
Accessories: None

Make: **Vortexion (Clarke and Smith)**
Model: VTN 30W
Output power: 30W
Frequency range: 30Hz-14kHz
Tone controls: Bass and treble
Inputs: 3
Outputs: 100V line; 8 ohms
Power supply: Mains
Weight: 7kg
Dimensions: 305 x 241 x 98mm
Cost: (b)
Accessories: None

Make: **Vortexion (Clarke and Smith)**
Model: VTN 50/70W
Output power: 50W
Frequency range: 30Hz-14kHz
Tone controls: Bass and treble
Inputs: 4 or 5
Outputs: 100V line; 8 ohms
Power supply: Mains
Weight: 10kg
Dimensions: 356 x 292 x 115mm
Cost: (c)
Accessories: None

RADIO CASSETTE RECORDERS

A radio cassette recorder consists of a radio receiver and cassette recorder built together. Since circuitry can be shared between the two functions the cost of the combined unit is generally less than a similar quality package of two separate devices. A second advantage arises when the recording of off-air programmes has to be carried out. It should be noted that this is only permitted when the programme being recorded is one produced for educational purposes.

Most of the units listed have a VHF waveband and some are stereo units. However care should be taken when interpreting the data since some listed as stereo can only replay stereo tapes and do not record in stereo from broadcast signals.

Since this type of equipment is extremely popular in the domestic market there are several hundred different models available but in the following list only those designed for use in the educational field have been included.

Make: **Clarke and Smith/JVC**
Model: RC-324LB
Wavebands: VHF, LW, MW, SW
Aerials: Telescopic and ferrite rod
Output power: 5W
Loudspeaker: 160mm, circular
Frequency range: nia
Tone controls: Treble
Recording level: Automatic
Microphone: Built-in
Monitor on record: nia
Inputs: Mic; line
Outputs: Line; earphone
Pause: nia

Cue and review:	nia
Tape counter:	3 digit
Power supply:	110; 220, 240V AC; 9V DC; batteries
Weight:	4kg
Dimensions:	243 x 355 x 110mm
Cost:	(a)
Accessories:	None

Make:	**Coomber**
Model:	308
Wavebands:	FM only
Aerials:	Telescopic
Output power:	4.5W
Loudspeaker:	200 x 75mm, elliptical
Frequency range:	nia
Tone controls:	Bass and treble
Recording level:	Automatic
Microphone:	External
Monitor on record:	nia
Inputs:	Mic; line record; line amplify
Outputs:	Line; ext speaker
Pause:	Yes
Cue and review:	No
Tape counter:	3 digit
Power supply:	Mains
Weight:	5.7kg
Dimensions:	178 x 362 x 328mm
Cost:	(b)
Accessories:	Microphone; headphones; ext loudspeaker
Other models:	310 – as above with 15W output

Make:	**Coomber**
Model:	342
Wavebands:	FM only
Aerials:	Telescopic
Output power:	4.5W
Loudspeaker:	254mm, circular
Frequency range:	nia
Tone controls:	Bass and treble
Recording level:	Automatic
Microphone:	External
Monitor on record:	nia
Inputs:	Mic; line record; line amplify
Outputs:	Line; ext speaker
Pause:	Yes
Cue and review:	No
Tape counter:	3 digit
Power supply:	Mains
Weight:	6kg
Dimensions:	380 x 337 x 210mm
Cost:	(b)
Accessories:	Microphone; headphones; ext speaker
Other models:	344 – as above with 15W output

Make:	**Farnell**
Model:	RC 15
Wavebands:	LW, MW, SW, FM
Aerials:	Telescopic
Output power:	15W per channel
Loudspeaker:	Elliptical
Frequency range:	nia
Tone controls:	Bass and treble
Recording level:	nia
Microphone:	nia
Monitor on record:	nia
Inputs:	Mic; line
Outputs:	Ext speaker
Pause:	nia
Cue and review:	Yes
Tape counter:	nia
Power supply:	Mains
Weight:	nia
Dimensions:	nia
Cost:	(b)
Accessories:	Ext speaker
Notes:	Stereo

Make:	**Goodsell**
Model:	CU.3
Wavebands:	VHF
Aerials:	Telescopic
Output power:	20W
Loudspeaker:	203mm, circular
Frequency range:	Amplifier, 50Hz-20kHz; tape deck, 100Hz-8kHz
Tone controls:	Bass and treble
Recording level:	Manual (metered)
Microphone:	External
Monitor on record:	No
Inputs:	Mic; line; gram
Outputs:	Line; external speaker
Pause:	Yes
Cue and review:	No
Tape counter:	3 digit
Power supply:	240V
Weight:	12kg
Dimensions:	450 x 300 x 300mm
Cost:	(c)
Accessories:	Dust cover; microphone; various connecting leads
Notes:	Memory counter can be fitted

Make:	**Goodsell**
Model:	PCR 585
Wavebands:	VHF
Aerials:	Telescopic or external
Output power:	5W
Loudspeaker:	200 x 125mm, elliptical
Frequency range:	100Hz-8kHz
Tone controls:	Yes
Recording level:	Manual (metered)
Microphone:	External
Monitor on record:	No
Inputs:	Mic; line; aerial
Outputs:	Line; external speaker
Pause:	Yes
Cue and review:	No
Tape counter:	3 digit
Power supply:	240V
Weight:	4.5kg
Dimensions:	330 x 300 x 180mm
Cost:	(b)
Accessories:	Dust cover; microphone; various connecting leads
Other models:	PCR 510 – as above with 250mm circular speaker (b). PCR 1285 – as above with 12W output (b). PCR 2010 – as above with 20W output and 250mm circular speaker (b)
Notes:	Memory counter can be fitted

Make:	**Hacker**
Model:	RPC1 Sovereign
Wavebands:	VHF, MW, LW
Aerials:	Telescopic and external
Output power:	1.2W
Loudspeaker:	203 x 127mm
Frequency range:	40Hz-10kHz
Tone controls:	Bass and treble
Recording level:	Automatic
Microphone:	External
Monitor on record:	Yes
Inputs:	Mic; line; aerial
Outputs:	Earphone
Pause:	Yes

Cue and review:	No
Tape counter:	No
Power supply:	220-240V
Weight:	5.7kg
Dimensions:	265 x 392 x 110mm
Cost:	(a)
Accessories:	None

Make:	**Marantz**
Model:	CR-1400
Wavebands:	VHF-AM/FM
Aerials:	nia
Output power:	nia
Loudspeaker:	90mm, circular
Frequency range:	nia
Tone controls:	No
Recording level:	Automatic/manual (metered)
Microphone:	Built-in
Monitor on record:	No
Inputs:	Mic; line; remote
Outputs:	Line; ext speaker; headphones
Pause:	Yes
Cue and review:	Yes
Tape counter:	3 digit
Power supply:	240V AC, 6V DC or batteries
Weight:	4.3kg
Dimensions:	310 x 197 x 95mm
Cost:	(a)
Accessories:	None
Other models:	CR 1403L – as above but VHF and LW (a). CR 1403S – as above but VHF and SW (a)
Notes:	Timer; PA facility

Make:	**Marantz**
Model:	CR 1200
Wavebands:	VHF-AM/FM
Aerials:	nia
Output power:	nia
Loudspeaker:	90mm, circular
Frequency range:	nia
Tone controls:	No
Recording level:	Automatic
Microphone:	Built-in
Monitor on record:	Yes
Inputs:	Mic; line; remote
Outputs:	Line; ext speaker; headphones
Pause:	No
Cue and review:	No
Tape counter:	No
Power supply:	240V AC or batteries
Weight:	3kg
Dimensions:	310 x 197 x 95mm
Cost:	(a)
Accessories:	None
Other models:	CR 1203L – as above but VHF and LW (a). CR 1203S – as above but VHF and SW (a)

Make:	**Marantz**
Model:	CR 1050
Wavebands:	VHF-AM/FM
Aerials:	nia
Output power:	nia
Loudspeaker:	Integral
Frequency range:	nia
Tone controls:	No
Recording level:	Automatic
Microphone:	Built-in
Monitor on record:	Yes
Inputs:	Mic; line
Outputs:	Line; ext speakers; earphone
Pause:	No
Cue and review:	No
Tape counter:	No
Power supply:	240V AC or batteries
Weight:	2.65kg
Dimensions:	280 x 182 x 83mm
Cost:	(a)
Accessories:	None
Other models:	CR 1053L – as above but VHF and LW (a). CR 1053S – as above but VHF and SW (a)

RADIO RECEIVERS

There are several hundred radio receivers now available on the domestic market, many of which are suitable for use in education or training, but the list which follows is made up from those designed specifically for this use. Many have the ability to receive signals from several wavebands but to enable the BBC Schools Broadcasts to be picked up it is essential to have a VHF waveband. When purchasing such a receiver the size of listening group should be kept in mind and a model with adequate audio output chosen.

Make:	**Clarke and Smith**
Model:	1170
Wavebands:	VHF, MW, LW
Aerials:	Telescopic and ferrite rod; external
Output power:	4W
Loudspeaker:	180 x 120mm, elliptical
Frequency range:	nia
Tone controls:	Treble
Inputs:	Line; aerials
Outputs:	Line; external speaker
Power supply:	220-250V
Weight:	3.3kg
Dimensions:	235 x 300 x 150mm
Cost:	(a)
Accessories:	None

Make:	**Coomber**
Model:	144
Wavebands:	VHF
Aerials:	Telescopic, built-in or external
Output power:	4.2W
Loudspeaker:	254mm, circular
Frequency range:	nia
Tone controls:	No
Inputs:	Aerial
Outputs:	Line
Power supply:	240V
Weight:	5.3kg
Dimensions:	305 x 470 x 150mm
Cost:	(a)
Accessories:	None
Other models:	146 – as above but mains/battery

Make:	**Coomber**
Model:	155
Wavebands:	VHF
Aerials:	Telescopic, built-in or external
Output power:	15W
Loudspeaker:	254mm, circular
Frequency range:	nia
Tone controls:	No
Inputs:	Line; aerial

Outputs:	Line; external speaker
Power supply:	240V
Weight:	5.7kg
Dimensions:	305 x 470 x 150mm
Cost:	(b)
Accessories:	None
Other models:	156 – as above with 70/100 volt line output

Make:	**Coomber**
Model:	157
Wavebands:	VHF
Aerials:	Telescopic, built-in or external
Output power:	30W
Loudspeaker:	254mm, circular
Frequency range:	nia
Tone controls:	No
Inputs:	Line; aerial
Outputs:	Line; external speaker
Power supply:	240V
Weight:	7kg
Dimensions:	305 x 470 x 150mm
Cost:	(b)
Accessories:	None

Make:	**Goodsell**
Model:	TR
Wavebands:	VHF
Aerials:	Telescopic; external
Output power:	5W
Loudspeaker:	200 x 125mm, elliptical
Frequency range:	nia
Tone controls:	nia
Inputs:	Aerial
Outputs:	Line; external speaker
Power supply:	240V
Weight:	4.5kg
Dimensions:	360 x 290 x 150mm
Cost:	(a)
Accessories:	None
Other models:	TRL – as above with 12W output (a). TRH – as TRL with 70/100 volt line output (a)

TELEVISION RECEIVER/MONITORS

A television receiver will not accept direct video or audio signals unless they have been converted to radio frequency via a modulator. To reproduce these direct signals two input sockets have to be provided – audio in and video in. Receiver/monitors are more expensive than their receiver only counterparts as the sockets have to be isolated from the mains supply, usually by the monitor having a built-in isolating mains transformer.

Make:	**Decca**
Model:	PC 6660 AL
Tube:	26in colour
Sound output power:	2.5W internal, 10W external loudspeaker
Sockets:	Aerial, video in and out 2 x BNCs, audio in and out 2 x DINs, ext loudspeaker DIN, remote control
Doors:	No
Trolley:	Optional extra
Power supply:	120-240V 50Hz
Weight:	nia
Dimensions:	550 x 525 x 815mm
Accessories:	None
Other models:	PZ 1000 – has full doors
Notes:	Lockable door over front controls

Make:	**Radio Rentals Contracts**
Model:	1765
Tube:	26in colour
Sound output power:	2.5W RMS; tone control
Sockets:	Aerial, video in and out BNCs and UHFs, audio in and out DINs, E1AJ 8 pin
Doors:	Lockable with shade above screen
Trolley:	Special Unicol available
Power supply:	240V 50Hz
Weight:	59kg
Dimensions:	550 x 790 x 581mm
Accessories:	Cordless remote control for all major functions
Other models:	1764
Notes:	Schools model. Touch timing with LED channel indicator. Teletext facilities may be added. Receiver/monitor switch on front panel

Make:	**Sony**
Model:	CUM 1100K Portable
Tube:	11in black and white
Sound output power:	nia
Sockets:	Aerial, video in and out UHFs, audio in and out 3.5mm jacks, E1AJ 8 pin
Doors:	No
Trolley:	No
Power supply:	240V 50Hz
Weight:	7.5kg
Dimensions:	286 x 295 x 302mm
Accessories:	nia

Make:	**Sony**
Model:	CUM 1810UB
Tube:	18in Trinitron colour PAL modified NTSC playback with suitable recorder
Sound output power:	1.5W
Sockets:	Aerial, video in and out UHFs, audio in and out 3.5mm jacks, E1AJ 8 pin monitor output – UHF and 3.5mm jack
Doors:	No
Trolley:	SV 1000
Power supply:	120/220/240V internal charge 50/60Hz
Weight:	34kg
Dimensions:	577 x 403 x 371mm

Make:	**Sony**
Model:	PVM 200CE
Tube:	20in black and white
Sound output power:	2W internal, 10W external loudspeaker
Sockets:	Video in UHF, audio in 3.5mm jack, E1AJ 8 pin, loudspeaker output
Doors:	No
Trolley:	No
Power supply:	100/120/240V 50/60Hz
Weight:	29kg
Dimensions:	515 x 536 x 370mm

Make:	**Tyne**
Model:	26in receiver/monitor with doors
Tube:	26in colour
Sound output power:	3W
Sockets:	Aerial, video in and out UHFs, audio in and out DINs, E1AJ 8 pin
Doors:	Yes
Trolley:	Optional
Power supply:	nia
Weight:	nia
Dimensions:	nia
Other models:	26in receiver/monitor without doors. 26in

receiver with VCR switch at rear (for use with VCRs which have an RF output), with doors. 26in receiver with VCR switch at rear, without doors. 22in receiver with VCR switch at rear, with doors. 22in receiver with VCR switch at rear, without doors

TAPE/VISUAL

DISSOLVE UNITS

When slides are being projected using a single projector, there is an inevitable one-and-a-half to two seconds during each slide change when the screen is totally dark. To prevent this, a system of using two projectors has been evolved. Each projector is used alternately and by overlapping the images from each on the screen it is not only possible to get rid of the dark period but many interesting and useful images are produced when both projectors are lit. This technique is known as 'fade and dissolve'. The automatic control of this technique, and of its logical extension to more than two projectors, is possible by using dissolve units.

A dissolve unit generates a varying frequency signal or a series of pulses of varying frequency. This signal can be decoded and can control the slide change of each projector and the brightness of each projector lamp. The control signal is recorded on the top quarter track of an audio tape. Some units have their own built-in cassette record and replay system while others require connection to a tape recorder or cassette recorder with a special facility for recording on tracks 1, 2 and 4 simultaneously. Cassette recorders of this type are listed under 'synchronising tape recorders'.

While the results of this technique are spectacular, care must be taken in purchasing the necessary equipment since different manufacturers have developed different control systems which are not compatible with each other. A second problem is the need for the projectors to have special control circuitry for both the slide change mechanism and the control of the lamp brightness. Some projectors have these built-in when they are manufactured, while others need to be modified to accommodate the particular system being used.

Make: **AV Equipment**
Model: AV 655 Duomaster
Control system: Present dissolve speed, 3-10 seconds; projector reset; external triggering signal required
Controls: Dissolve rate
Hand control: None
Connections: Flying leads for projectors
Power supply: 220/240V
Weight: 2kg
Dimensions: 250 x 80 x 120mm
Cost: (c)
Accessories: None

Make: **AV Equipment**
Model: AV 4000 Audio Convar
Control system: Dissolve control 1500-3000Hz, slide changes, A 3750Hz, B 750Hz
Controls: Microphone and line mixers; volume
Hand control: Slider for dissolve and slide changes
Connections: Hand control; flying leads for projectors
Power supply: 220/240V
Weight: 4.5kg
Dimensions: 300 x 85 x 240mm
Cost: (f)
Accessories: Carrying case with integral loud speakers
Notes: Built-in cassette unit; power amplifier, 15W; stereo playback using external amplifiers

Make: **AV Equipment**
Model: AV 4004 Pro Convar
Control system: Dissolve control 1500-3000Hz, slide changes, A 3750Hz, B 750Hz
Controls: Playback
Hand control: Slider for dissolve and slide changes
Connections: Hand control and tape recorder sockets; flying leads for projectors
Power supply: 220/240V
Weight: 5.0kg
Dimensions: 250 x 76 x 250mm
Cost: (b)
Accessories: None
Notes: Tape recorder required

Make: **AV Equipment**
Model: AV 4300 Audiomatic
Control system: Dissolve control 1500-3000Hz, slide changes, A 3750Hz, B 750Hz
Controls: Record/playback; volume; tone
Hand control: Slider for dissolve and slide changes
Connections: Hand control socket; flying leads for projectors
Power supply: 220/240V
Weight: 4.5kg
Dimensions: 300 x 80 x 240mm
Cost: (e)
Accessories: None
Notes: Cartridge sound system; 15W amplifier

*

Make: **Edric**
Model: Edrimatic 500 M
Control system: nia
Controls: Record/replay; reset
Hand control: Slide change; dissolve; cut; superimpose
Connections: Hand control; projectors; tape recorder
Power supply: Mains
Weight: nia
Dimensions: nia
Cost: (d)
Accessories: None
Other models: 501M – as above but up to 10 projectors can be controlled

*see late entry page 94

Notes: Requires connection to a suitable tape recorder

Make: **Electrosonic**
Model: Showslide ES 3069
Control system: Variable frequency
Controls: Line up; mains
Hand control: Record/play; slider for dissolve; flip
Connections: Hand control; tape recorder; flying leads to projectors
Power supply: Mains
Weight: 2.6kg
Dimensions: 260 x 80 x 220mm
Cost: (c)
Accessories: None
Notes: Requires suitable tape recorder for auto control

Make: **Electrosonic**
Model: Showpulse ES 3006
Control system: 1000Hz control signal; cut 370ms, dissolve 65ms; dissolve rate pre-set 3-12 seconds
Controls: Line up; record/play; timer; mains
Hand control: Cut; dissolve; reverse
Connections: Hand control; tape recorder; projectors
Power supply: Mains
Weight: 2.6kg
Dimensions: 260 x 80 x 220mm
Cost: (c)
Accessories: None
Notes: Requires suitable tape recorder for auto control

Make: **Imatronic**
Model: Digital SX Manual
Control system: Digital
Controls: Twin sliding manual controls
Hand control: None
Connections: Leads to projectors
Power supply: 24V, 50Hz (from projectors)
Weight: 0.5kg
Dimensions: 195 x 95 x 35mm
Cost: (b)
Accessories: None

Make: **Imatronic**
Model: Digital SX 2500
Control system: Digital pulse counting-carrier frequency 2.5kHz
Controls: Record/playback
Hand control: (i) Slider for dissolve and slide advance, push-button flash; (ii) buttons for slide advance, superimpose, fast/slow dissolve, dissolve rate control
Connections: Tape recorder; hand control; flying leads to projector
Power supply: 24V 50Hz (from projectors)
Weight: 1.5kg
Dimensions: 260 x 230 x 70mm
Cost: (c)
Accessories: Hand control
Other models: SX 2500E – as above with built-in electronic hand control (c)

Make: **Kodak**
Model: Model B
Control system: (i) Impulse mode – 1kHz pulses control pre-set dissolve timer and slide changer; (ii) frequency modulation mode – hand control slider controls rate of dissolve and slide change
Controls: Dissolve time; interval; address select; mains switch
Hand control: (i) Impulse mode; (ii) frequency modulation mode
Connections: Projectors; hand controls; tape recorder; mains
Power supply: 110 to 250V
Weight: 5.3kg
Dimensions: 325 x 75 x 290mm
Cost: nia
Accessories: None
Notes: Requires connection to a suitable tape recorder

Make: **Kodak**
Model: AV Presentation Unit
Control system: (i) Impulse mode – controls pre-set dissolve times and slide changes; (ii) digital mode – hand control slider controls rate of dissolve and slide change
Controls: Record/playback; Dolby; impulse/digital; timer; mains
Hand control: (i) Impulse mode; (ii) digital mode
Connections: Audio in; hand controls; mains
Power supply: 110/130/220-250V
Weight: Main unit is 18kg; speakers; 7.2kg
Dimensions: Main unit (closed) 850 x 400 x 200mm; speakers (closed) 380 x 360 x 216mm
Cost: (k)
Accessories: None
Notes: Unit is complete with twin extension speakers; power amplifier; PA facility; remote control facility; requires two S-AV 2020 projectors

Make: **Leitz**
Model: DU-24A
Control system: nia
Controls: Line up; hand control
Hand control: Dissolve; slide change; flip; erase, play and record
Connections: Hand control; tape recorder; free lead
Power supply: Mains
Weight: nia
Dimensions: nia
Cost: nia
Accessories: None
Notes: Requires connection to a suitable tape recorder

Make: **Leitz**
Model: DU-24M
Control system: Manual control only; forward and reverse slide change; dissolve time pre-set 11 to 30 secs
Controls: nia
Hand control: Forward; reverse; timer; hard/soft cut
Connections: Flying leads to projectors
Power supply: Connects directly to projectors
Weight: nia
Dimensions: nia
Cost: nia
Accessories: None

FILMSTRIP/SOUND VIEWERS

The equipment in this category is mostly designed to operate with the visual component of the programme in the form of a 35mm filmstrip of the single frame format. The audio signal, and control pulses, are recorded on a Compact Cassette. All models have playback facilities while some can also be used for recording purposes. Since all models have a rear projection screen they are classed as viewers but some can be used to project on to a front projection screen.

The most common use of these machines is in the promotional field and they are not often used in either education or training.

A second system can be used in this category and this is featured in the equipment marketed by Fairchild. Here the filmstrip is 16mm and the control signals are more complex. This enables the user to select parts of the programme and the internal circuitry ensures that both the audio and visual components are always synchronised.

Make:	**Avcom**
Model:	2500
Filmstrip size:	200 frames/35mm half frame
Magazine type:	Special cartridge
Screen size:	Projector only
Lamp:	240V, 300W
Lens:	75mm
Optional lenses:	35mm; 50mm; 100mm; 135mm
Tape:	Compact cassette
Synchronising system:	1000Hz advance, track 4
Output power:	2W
Loudspeaker:	80 x 120mm, elliptical
Frequency range:	nia
Tone controls:	No
Outputs:	nia
Power supply:	Mains
Weight:	3.5kg
Dimensions:	373 x 215 x 110mm
Cost:	(d)
Accessories:	Brief-case with ext speaker; remote control; microphone
Other models:	5000 – similar to above but 150 frames/35mm full frame (d)

Make:	**Bell and Howell**
Model:	Filmosound 35-767
Filmstrip size:	156 frames; 35mm single frame
Magazine type:	Special combined film and tape cartridge
Screen size:	248 x 187mm
Lamp:	nia
Lens:	nia
Tape:	Compact cassette
Synchronising system:	50Hz and 1000Hz, track 4
Output power:	nia
Loudspeaker:	Integral
Frequency range:	nia
Tone controls:	No
Outputs:	Headphones
Power supply:	110/220/240V
Weight:	9kg
Dimensions:	nia
Cost:	(c)
Accessories:	Remote hand switch; remote fast switch; headphones; dual headphone; junction box
Other models:	Filmosound 35-768, portable version of 767 (c)
Notes:	Fast forward and reverse search

Make:	**Dukane**
Model:	A-V Matic – 28 A2A
Filmstrip size:	35mm single frame
Magazine type:	None
Screen size:	267 x 203mm
Lamp:	nia
Lens:	nia
Tape:	Compact cassette
Synchronising system:	nia
Output power:	nia
Loudspeaker:	Integral, 75 x 125mm, elliptical
Frequency range:	nia
Tone controls:	nia
Outputs:	Headphones
Power supply:	120 or 240V
Weight:	8.2kg
Dimensions:	280 x 330 x 457mm
Cost:	(d)
Accessories:	Headphones; remote control; cover; 4 headphone distribution box
Notes:	Pause control; automatic threading

Make:	**Dukane**
Model:	Micromatic II 28A82
Filmstrip size:	35mm half frame (50 x 50mm slides with adaptor)
Magazine type:	None
Screen size:	Projector only
Lamp:	240V, 300 or 500W
Lens:	76mm f/2.5
Optional lenses:	38mm f/3; 127mm f/2.8
Tape:	Compact cassette
Synchronising system:	50Hz and 1000Hz pulses
Output power:	nia
Loudspeaker:	Integral
Frequency range:	nia
Tone controls:	No
Outputs:	nia
Power supply:	240V
Weight:	8.9kg
Dimensions:	394 x 445 x 184mm
Cost:	(d)
Accessories:	Ext speaker; dust cover; remote control; manual slide changer; stack loader
Notes:	Pause control; two speeds for advance and rewind

Make:	**Edric**
Model:	Audiscan 4000
Filmstrip size:	350 frames/16mm
Magazine type:	Combined film/tape
Screen size:	140 x 190mm
Lamp:	Low voltage
Lens:	nia
Tape:	¼in
Synchronising system:	nia
Output power:	nia
Loudspeaker:	Integral
Frequency range:	nia
Tone controls:	No
Outputs:	Headphones
Power supply:	Mains
Weight:	10kg
Dimensions:	457 x 305 x 190mm
Cost:	(d)
Accessories:	Dust cover; transit case
Other models:	4200 – as above with fast forward in

	sync facility (d). 3200 – as 4200 with larger screen (e)
Notes:	Front and rear projection; economy lamp switch

Make:	**Elf**
Model:	Audio Vu Cassette SP
Filmstrip size:	150 frames/35mm half frame
Magazine type:	Auto spool
Screen size:	Projector only
Lamp:	240V, 300W
Lens:	85mm f/2.8
Tape:	Compact cassette
Synchronising system:	1000Hz advance, track 4
Output power:	1.5W
Loudspeaker:	105mm diameter
Frequency range:	nia
Tone controls:	No
Outputs:	Ext speaker; remote control
Power supply:	220/240V
Weight:	5.4kg
Dimensions:	144 x 325 x 230mm
Cost:	(d)
Accessories:	Remote control; conversion lens; dust cover
Notes:	Mic input; manual frame advance; power rewind

Make:	**Elmo**
Model:	35-FT
Filmstrip size:	200 frames/35mm half frame
Magazine type:	None
Screen size:	Projector only
Lamp:	240V, 300W
Lens:	85mm f/2.8
Tape:	Compact cassette
Synchronising system:	1000Hz advance, track 4
Output power:	3W
Loudspeaker:	Integral 100 x 20mm, elliptical
Frequency range:	nia
Tone controls:	No
Outputs:	nia
Power supply:	240V
Weight:	5kg
Dimensions:	300 x 275 x 115mm
Cost:	(c)
Accessories:	None

Make:	**La Belle**
Model:	Duo – 16 projector
Filmstrip size:	250 frames/16mm
Magazine type:	Commpak film/tape combined
Screen size:	130 x 175mm
Lamp:	Low voltage
Lens:	24mm f/1.6
Tape:	¼in
Synchronising system:	nia
Output power:	nia
Loudspeaker:	102mm diameter
Frequency range:	nia
Tone controls:	No
Outputs:	Headphones; ext speaker; remote control
Power supply:	nia
Weight:	8.2kg
Dimensions:	419 x 356 x 172mm
Cost:	(d)
Accessories:	Dust cover; remote control
Notes:	Front and rear projection

Make:	**La Belle**
Model:	Tutor 16
Filmstrip size:	250 frames/16mm
Magazine type:	Commpak filn/tape combined
Screen size:	Projector only
Lamp:	nia
Lens:	38mm f/1.6
Tape:	¼in
Synchronising system:	nia
Output power:	nia
Loudspeaker:	nia
Frequency range:	nia
Tone controls:	nia
Outputs:	nia
Power supply:	nia
Weight:	nia
Dimensions:	nia
Cost:	nia
Accessories:	None
Notes:	Front projection only; economy lamp switch

Make:	**La Belle**
Model:	Showman 16
Filmstrip size:	16mm
Magazine type:	Commpak combined sound/filmstrip
Screen size:	205 x 273mm
Lamp:	Low voltage
Lens:	nia
Tape:	¼in, 17/8in ips or 3¾in ips
Synchronising system:	Stop and advance
Output power:	nia
Loudspeaker:	Integral
Frequency range:	nia
Tone controls:	No
Outputs:	Remote control; ext speaker
Power supply:	nia
Weight:	8.2kg
Dimensions:	381 x 318 x 337mm
Cost:	(d)
Accessories:	Dust cover; pedestal; timer; counter; remote control
Other models:	La Belle produce several other models but no relevant information is available

Make:	**Simda**
Model:	VAL 77
Filmstrip size:	150 frames/35mm double frame
Magazine type:	Simda cartridge
Screen size:	160 x 230mm
Lamp:	12V, 20W
Lens:	Retrofocus, anastigmatic
Tape:	Compact cassette
Synchronising system:	1000Hz, track 4
Output power:	1W
Loudspeaker:	Integral
Frequency range:	60 to 12000Hz
Tone controls:	Treble
Outputs:	nia
Power supply:	110-220V AC or internal NiCd batteries
Weight:	16kg without batteries, 8kg with batteries
Dimensions:	450 x 350 x 130mm
Cost:	(i)
Accessories:	None
Notes:	Portable, attache case style, unit

Make:	**Singer**
Model:	Auto-Vance III
Filmstrip size:	35mm
Magazine type:	None
Screen size:	180 x 135mm
Lamp:	Low voltage
Lens:	40mm

Tape: Compact cassette
Synchronising system: 50Hz, audio track
Output power: nia
Loudspeaker: Integral
Frequency range: 150-6000Hz
Tone controls: No
Outputs: Earphone jack
Power supply: 117V
Weight: 5.7kg
Dimensions: 190 x 248 x 343mm
Cost: nia
Accessories: None
Other models: Auto Vance III SP – as above with remote control

Make: **Singer**
Model: Insta-Load 35
Filmstrip size: 100 or 250 frames, 35mm
Magazine type: None
Screen size: 190 x 250mm
Lamp: Low voltage
Lens: 28mm, and 60mm f/3
Tape: Compact cassette
Synchronising system: 50Hz and 1000Hz
Output power: 5W
Loudspeaker: Integral, 75 x 125mm, elliptical
Frequency range: 200-7000Hz
Tone controls: Treble
Outputs: nia
Power supply: 220/240V
Weight: 7.7kg
Dimensions: 457 x 229 x 330mm
Cost: nia
Accessories: None
Other models: Insta-Load 35 Cue-Lok – as above with search and review and micro-processor sync control; 50H2 pulses only

SLIDE/TAPE VIEWERS

The units listed below are primarily intended for individual or small group viewing of synchronised tape/slide sequences. The visual component will be in the form of standard 50 x 50mm slides while the audio will be recorded on a compact cassette. The cassette will also have the control information recorded using the separate track system.

Some models have facilities for recording both audio and control tracks while other models are playback only. Two types of control pulse can be recorded. All machines will respond to a slide change pulse while some models have a stop facility which can stop the tape, and hence the entire presentation, until restarted by the student operating the appropriate control.

All models have a built-in back projection screen for use in the individual or small group situation, but a few can also be used as projectors on to a normal front projection screen by opening a door in the cabinet.

Make: **Bell and Howell**
Model: Ring master
Slide size: 50 x 50mm
Magazine type: Carousel rotary (80)
Screen size: 240 x 240mm
Lamp: 19V, 80W
Lens: 75mm f/3.5
Optional lenses: None
Tape: Compact cassette
Synchronising system: 1000Hz advance, track 4
Output power: nia
Loudspeaker: Integral
Frequency range: nia
Tone controls: No
Outputs: Headphones; ext speaker
Power supply: 110/130/150/220/240V
Weight: 11kg
Dimensions: 290 x 370 x 360mm
Cost: (e)
Accessories: Carrying case; 'point of sale' cover
Other models: Record/replay model available
Notes: 'Frame filler' enlarges projected image; autofocus; front and rear projection
*

Make: **Fairchild**
Model: Syncroslide 35-3501
Slide size: 50 x 50mm
Magazine type: Carousel rotary (80)
Screen size: 230 x 230mm
Lamp: 19V, 80W
Lens: 63.5mm f/3.5
Optional lenses: None
Tape: Compact cassette
Synchronising system: 1000Hz advance, track 4
Output power: 2W
Loudspeaker: Integral, 60 x 150mm, elliptical
Frequency range: nia
Tone controls: No
Outputs: nia

*See late entry page 94

Power supply:	220-240V
Weight:	10.8kg
Dimensions:	406 x 315 x 333mm
Cost:	(d)
Accessories:	'Point of sale' cover; dust cover; carrying case; remote control
Other models:	3501 PBS – as above with auto programme stop (e); 3502 RPBS – as above with record facility (e); 3501 CCP – as 3501 with computerised control (f); 3505 PB – portable front projection version of 3501 (e); 3505 RPBS – recording version of 3505 PB (f)

Make:	**Kindermann**
Model:	AV 1002
Slide size:	50 x 50mm
Magazine type:	Leitz, straight (36 or 50)
Screen size:	180 x 180mm
Lamp:	nia
Lens:	nia
Optional lenses:	nia
Tape:	Compact cassette
Synchronising system:	Yes
Output power:	nia
Loudspeaker:	Built-in lid
Frequency range:	nia
Tone controls:	nia
Outputs:	Ext speaker
Power supply:	Mains
Weight:	nia
Dimensions:	nia
Cost:	(h)
Accessories:	None

Make:	**Kodak**
Model:	Cinemasound 4000
Slide size:	50 x 50mm
Magazine type:	Carousel rotary (80)
Screen size:	nia
Lamp:	300W
Lens:	nia
Optional lenses:	Yes
Tape:	Compact cassette
Synchronising system:	Built-in
Output power:	16W
Loudspeaker:	Side mounted
Frequency range:	50-10,000Hz
Tone controls:	No
Outputs:	Speaker; headphones; ext amplifier
Power supply:	nia
Weight:	11kg
Dimensions:	380 x 305 x 432mm
Cost:	nia
Accessories:	None
Notes:	Front and rear projection

Make:	**Kodak**
Model:	Cinemasound 5000
Soide size:	50 x 50mm
Magazine type:	Carousel rotary (80)
Screen size:	330 x 457mm
Lamp:	300W
Lens:	nia
Optional lenses:	Yes
Tape:	Compact cassette
Synchronising system:	Built-in
Output power:	16W
Loudspeaker:	Front mounted, 250mm
Frequency range:	50-10,000Hz
Tone controls:	No
Outputs:	Speaker; headphones; ext amplifier
Power supply:	nia
Weight:	17.3kg
Dimensions:	508 x 508 x 635mm
Cost:	nia
Accessories:	None
Notes:	Front and rear projection

Make:	**Singer**
Model:	Caramate 3300
Slide size:	50 x 50mm
Magazine type:	Carousel rotary (80)
Screen size:	230 x 230mm
Lamp:	19V, 80W
Lens:	77mm f/3.5 and 56mm f/3
Optional lenses:	No
Tape:	Compact cassette
Synchronising system:	150Hz pause, 1000H advance, track 4
Output power:	5W
Loudspeaker:	75 x 120mm, elliptical
Frequency range:	nia
Tone controls:	Treble
Outputs:	Earphone; ext speaker; remote control
Power supply:	220/240V
Weight:	9.5kg
Dimensions:	376 x 330 x 330mm
Cost:	(e)
Accessories:	None
Other models:	Record/playback model (f)
Notes:	Front and rear projection

Make:	**Technicolor**
Model:	Road Show 40E
Slide size:	50 x 50mm
Magazine type:	Carousel rotary (80)
Screen size:	260 x 260mm
Lamp:	120V, 300W
Lens:	76mm f/3.5
Optional lenses:	64mm f/3.5; 102mm f/2.8
Tape:	Compact cassette
Synchronising system:	1000Hz advance, track 4
Output power:	2W
Loudspeaker:	57 x 133mm, elliptical
Frequency range:	nia
Tone controls:	No
Outputs:	Headphone; remote control
Power supply:	220/240V
Weight:	11.8kg
Dimensions:	210 x 337 x 486mm
Cost:	(e)
Accessories:	Remote control
Other models:	Road Show 20E – as above without record facility
Notes:	Front and rear projection

SYNCHRONISING CASSETTE RECORDERS

These machines are designed to enable the visual part of an audio-visual presentation to be controlled by the tape carrying the audio signal. This can be accomplished in two distinct ways.

1. In the 'superimposed' system a control signal is recorded on the same track as the audio signal but at a frequency which is inaudible. On playback this signal is filtered off and used to operate the control devices on the slide projector(s).

2. In the 'separate track' system the recorder has an

additional record/playback head which is used for the control signal. Some recorders require additional equipment for supplying the control signal while others have the required circuitry built in. The control signal is recorded on the upper quarter of the tape while the audio signal is recorded on the lower half of the tape as is normal. This means that the audio signal can be mono or stereo without affecting the control function but it also restricts the use of the tape since no recordings can be made on the second track of the tape without erasing the control signal. There are two control signals in common use: a 1kHz pulse which signals a slide change and a 150Hz pulse which is used to stop the tape. The tape is restarted manually by the operator. Some of the separate track recorders have a socket which enables external signals to be recorded on the control track. These can then be used, with the appropriate external circuitry, to control fade and dissolve presentations using two or more projectors.

Make:	**Gnome**
Model:	Jonan 440 SL
Mono/stereo:	Mono
Tracks:	2
Synchronising system:	1kHz advance on track 2
Output power:	1.3W
Loudspeaker:	75 x 125mm, elliptical
Frequency range:	80Hz-8kHz
Tone controls:	Treble cut
Recording level:	Automatic/manual (metered)
Microphone:	External
Monitor on record:	No
Inputs:	Mic; line
Outputs:	Line; ext speaker
Pause:	No
Cue and review:	No
Tape counter:	3 digit
Power supply:	117-220V AC or 6V DC
Weight:	3.8kg
Dimensions:	230 x 224 x 73mm
Cost:	(a)
Accessories:	None

Make:	**Goodsell**
Model:	CTR 1485
Mono/stereo:	Mono
Tracks:	2
Synchronising system:	150Hz stop and 1kHz advance on track 2
Output power:	12W
Loudspeaker:	200 x 125mm, elliptical
Frequency range:	Amplifier 50Hz-20kHz, tape deck 100Hz-8kHz
Tone controls:	Bass and treble
Recording level:	Manual (metered)
Microphone:	External
Monitor on record:	No
Inputs:	Mic; line; sync
Outputs:	Line; sync; projector
Pause:	Yes
Cue and review:	No
Tape counter:	3 digit
Power supply:	240V
Weight:	5kg
Dimensions:	330 x 300 x 180mm
Cost:	(c)
Accessories:	Dust cover; microphone; various connecting leads
Other models:	CTP 2310 – as above with 20W output; 250mm circular speaker; facilities for external sync and dissolve units (c)
Notes:	Memory counter can be fitted

Make:	**ITT**
Model:	740 AV
Mono/stereo:	Stereo
Tracks:	2 plus control
Synchronising system:	Free recording head on track 4
Output power:	1.5W
Loudspeaker:	Integral
Frequency range:	40Hz-12.5kHz
Tone controls:	No
Recording level:	Manual (metered)
Microphone:	External
Monitor on record:	Yes
Inputs:	Mic; line; (separately controlled and with mixer)
Outputs:	Line; headphones; ext speaker
Pause:	Yes
Cue and review:	No
Tape counter:	3 digit
Power supply:	200-240V or 7.5V DC
Weight:	3.5kg
Dimensions:	315 x 85 x 270mm
Cost:	(d)
Accessories:	Headphones; microphone; synchroniser; power pack; charger

Make:	**Philips**
Model:	N 2229
Mono/stereo:	Mono
Tracks:	1 plus 1
Synchronising system:	1kHz advance on track 2 using external synchroniser
Output power:	1.5W
Loudspeaker:	Integral
Frequency range:	nia
Tone controls:	Treble cut, switched
Recording level:	Automatic/manual
Microphone:	Built-in
Monitor on record:	Yes
Inputs:	Mic; line; sync
Outputs:	Line; sync; ext speaker
Pause:	Yes
Cue and review:	No
Tape counter:	3 digit
Power supply:	Mains/battery
Weight:	nia
Dimensions:	250 x 80 x 275mm
Cost:	(a)
Accessories:	Sync control

Make:	**Rank Aldis**
Model:	152 S
Mono/stereo:	Mono
Tracks:	2
Synchronising system:	1kHz advance on track 2 or 50Hz on audio track
Output power:	2W
Loudspeaker:	150mm, elliptical
Frequency range:	40Hz-10kHz
Tone controls:	Treble cut
Recording level:	Automatic/manual
Microphone:	Built-in/external
Monitor on record:	No
Inputs:	Mic; line
Outputs:	Headphones; sync
Pause:	Yes
Cue and review:	Yes
Tape counter:	3 digit
Power supply:	240V
Weight:	2.6kg
Dimensions:	254 x 235 x 90mm
Cost:	(b)

Accessories:	Microphones; headphones; sync cord

Make:	**Rank Aldis**
Model:	1625-2
Mono/stereo:	Mono
Tracks:	2
Synchronising system:	1kHz advance, 150Hz stop on track 2 or 50Hz on audio track
Output power:	10W
Loudspeaker:	250mm, elliptical
Frequency range:	40Hz-10kHz
Tone controls:	Treble cut
Recording level:	Automatic/manual
Microphone:	Built-in/external
Monitor on record:	No
Inputs:	Mic; line; sync
Outputs:	Headphones; sync
Pause:	Yes
Cue and review:	Yes
Tape counter:	3 digit
Power supply:	240V
Weight:	6.1kg
Dimensions:	356 x 254 x 165mm
Cost:	(d)
Accessories:	Microphones; headphone; sync cord

Make:	**Telex (Avcom Systems Ltd)**
Model:	C-130
Mono/stereo:	Mono
Tracks:	2
Synchronising system:	150Hz stop, 1000Hz advance, track 2
Output power:	10W
Loudspeaker:	Integral
Frequency range:	40Hz-10kHz
Tone controls:	Yes
Recording level:	Manual (metered)
Microphone:	External
Monitor on record:	No
Inputs:	Mic; line; sync; remote
Outputs:	Line; sync; ext speaker
Pause:	Yes
Cue and review:	Yes
Tape counter:	3 digit
Power supply:	220V
Weight:	7.2kg
Dimensions:	133 x 254 x 374mm
Cost:	(d)
Accessories:	Foot switch; filmstrip sync cable
Other models:	C-140 – replay only version of C-130 (c)

Make:	**Wollensak (3M)**
Model:	2551 AV
Mono/stereo:	Mono
Tracks:	2
Synchronising system:	1kHz advance on track 2
Output power:	9W
Loudspeaker:	Integral
Frequency range:	50Hz-10kHz
Tone controls:	Treble cut
Recording level:	Automatic/manual (metered)
Microphone:	External (supplied)
Monitor on record:	No
Inputs:	Mic; line; sync
Outputs:	Line; sync; ext speaker
Pause:	Yes
Cue and review:	No
Tape counter:	3 digit
Power supply:	120-240V
Weight:	8.172kg
Dimensions:	314 x 365 x 165mm
Cost:	(e)
Accessories:	None

VIDEO

VIDEO CASSETTE RECORDERS

The following list of video cassette recorders (VCR) is by no means an exhaustive one. To compile such a list would be an endless task since new models come on to the market at a slightly faster rate than the old models are deleted. Most of the cheaper machines listed are primarily intended for the domestic market but this does not preclude their use in either education or training.

There are several different types of tape cassette in common use and these are not interchangeable. Generally, Japanese manufacturers use the three systems designated VHS, Betamax and U-matic while their European counterparts use VCR, VCR LP and VCR 2000. A new type of tape is now making an appearance which uses ¼ inch tape in a cassette similar to, but slightly larger than, the standard audio cassette.

Where a maximum playing time is indicated this refers to the manufacturers' own recommended tape. It may be that a tape produced by another firm will be compatible with the VCR and give a longer playing time.

Many of the domestic recorders, other than portable models, have built-in tuner circuits which enable them to record directly off air. The provision of timers on many machines is increasing as is the complexity of the control functions which they afford.

An audio dubbing facility indicates that it is possible to record an audio track after the video signal has been recorded. This makes it possible to add your own commentary and/or sound effects after producing the visual part of the programme.

Electronic editing enables the video signal to be added to or altered. Two types of editing are possible:

1. *Assemble edit.* In this case a new piece of video material can be recorded after a previous one without break-up of the picture.
2. *Insert edit.* This permits the insertion of new material in to an already completed sequence with-

out the picture breaking up either before or after the edit.

Portable models are available which usually obtain their power from built-in rechargeable batteries, a car battery or the mains power supply.

Make:	**Akai**
Model:	VP-7100
Tape:	VHS
Max running time:	180min
Recording level:	Automatic
Still picture:	No
Slow motion:	No
Tuner:	No
Timer:	No
Audio dub:	Yes
Electronic edit:	No
Tape counter:	3 digit
Inputs:	Video; audio (mic and line)
Outputs:	Video; audio (line)
Power supply:	12V DC
Weight:	7.5kg
Dimensions:	388 x 137 x 328mm
Cost:	(i)
Accessories:	Remote pause tuner; AC adaptor

Make:	**Akai**
Model:	VS-9700
Tape:	VHS
Max running time:	180min
Recording level:	Automatic
Still picture:	No
Slow motion:	No
Tuner:	12 channel
Timer:	8 day
Audio dub:	Yes
Electronic edit:	No
Tape counter:	3 digit
Inputs:	Video; aerial; audio (mic and line)
Outputs:	Video; aerial; audio (line)
Power supply:	110/220/240V
Weight:	14.5kg
Dimensions:	480 x 153 x 334mm
Cost:	nia
Accessories:	None

Make:	**Akai**
Model:	VS-9800
Tape:	VHS
Max running time:	180min
Recording level:	Automatic
Still picture:	Yes
Slow motion:	Yes
Tuner:	8 channel
Timer:	8 day
Audio dub:	Yes
Electronic edit:	No
Tape counter:	3 digit
Inputs:	Video; aerial; audio (mic and line)
Outputs:	Video; aerial; audio (line)
Power supply:	110/127/220/240V
Weight:	14kg
Dimensions:	451 x 147 x 352mm
Cost:	(h)
Accessories:	None

Make:	**Bush**
Model:	BV 6900
Tape:	Betamax
Max running time:	195min
Recording level:	Automatic
Still picture:	Yes

Slow motion:	Yes
Tuner:	10 channel
Timer:	7 day/3 programme
Audio dub:	Yes
Electronic edit:	Assemble
Tape counter:	4 digit
Inputs:	Video; aerial; audio (mic and line)
Outputs:	Video; aerial; audio (line)
Power supply:	240V
Weight:	13.5kg
Dimensions:	475 x 178 x 386mm
Cost:	(f)
Accessories:	None

Make:	**Ferguson**
Model:	3V01
Tape:	VHS
Max running time:	180min
Recording level:	Automatic
Still picture:	Yes
Slow motion:	No
Tuner:	No
Timer:	No
Audio dub:	Yes
Electronic edit:	Assemble
Tape counter:	3 digit
Inputs:	Video; audio (mic and line)
Outputs:	Video; audio (line)
Power supply:	12V DC
Weight:	7.5kg
Dimensions:	338 x 137 x 328mm
Cost:	nia
Accessories:	Timer/tuner; AC adaptor

Make:	**Ferguson**
Model:	3V22
Tape:	VHS
Max running time:	180min
Recording level:	Automatic
Still picture:	No
Slow motion:	No
Tuner:	8 channel
Timer:	7 day/1 programme
Audio dub:	Yes
Electronic edit:	Assemble
Tape counter:	3 digit
Inputs:	Video; aerial; audio (mic and line)
Outputs:	Video; aerial; audio (line)
Power supply:	240V
Weight:	14kg
Dimensions:	453 x 147 x 337mm
Cost:	nia
Accessories:	None
Other models:	3V16 – as above with still picture, slow motion and remote control

Make:	**Ferguson**
Model:	3V23
Tape:	VHS
Max running time:	180min
Recording level:	Automatic
Still picture:	Yes
Slow motion:	Yes
Tuner:	16 channel
Timer:	14 day/8 programme
Audio dub:	Yes
Electronic edit:	No
Tape counter:	4 digit
Inputs:	Video; aerial; audio (mic and line)
Outputs:	Video; aerial; audio (line)
Power supply:	240V
Weight:	16.5kg
Dimensions:	470 x 153 x 387mm
Cost:	nia
Accessories:	None
Notes:	Infrared remote control supplied

Make:	**Ferguson**
Model:	3V24
Tape:	VHS
Max running time:	180min
Recording level:	Automatic
Still picture:	Yes
Slow motion:	No
Tuner:	No
Timer:	No
Audio dub:	Yes
Electronic edit:	Assemble
Tape counter:	3 digit
Inputs:	Video; audio (mic and line)
Outputs:	Video; audio (line)
Power supply:	12V DC
Weight:	4.4kg
Dimensions:	288 x 103 x 267mm
Cost:	nia
Accessories:	Tuner/timer; AC adaptor

Make:	**Grundig**
Model:	VCR 601
Tape:	VCR
Max running time:	60min
Recording level:	Automatic/manual
Still picture:	Yes
Slow motion:	Yes
Tuner:	No
Timer:	No
Audio dub:	Yes
Electronic edit:	Assemble/insert
Tape counter:	Elapsed time counter
Inputs:	Video; audio (mic and line)
Outputs:	Video; audio (line)
Power supply:	220V AC or 12V DC
Weight:	10kg
Dimensions:	446 x 152 x 373mm
Cost:	(k)
Accessories:	None

Make:	**Grundig**
Model:	Video 2xf
Tape:	Video compact cassette
Max running time:	2 x 240min
Recording level:	Automatic
Still picture:	No
Slow motion:	No
Tuner:	8 channel
Timer:	10 day/4 programme
Audio dub:	Yes
Electronic edit:	No
Tape counter:	3 digit
Inputs:	Aerial; audio (mic)
Outputs:	Aerial
Power supply:	220V
Weight:	14.5kg
Dimensions:	590 x 170 x 320mm
Cost:	(g)
Accessories:	Remote control
Other models:	Video 2 x 4 Plus – as above with still picture and slow motion (g)

Make:	**Hitachi**
Model:	SV-340
Tape:	U
Max running time:	20min
Recording level:	Video/auto; audio/manual or auto
Still picture:	Yes

Slow motion:	No
Tuner:	No
Timer:	No
Audio dub:	Yes
Electronic edit:	Assemble
Tape counter:	3 digit
Inputs:	Video; audio (mic and line)
Outputs:	Video; audio (line)
Power supply:	12V DC
Weight:	11.2kg
Dimensions:	336 x 141 x 369mm
Cost:	nia
Accessories:	AC adaptor; battery pack; RF converter

Make:	**Hitachi**
Model:	VT-5500E
Tape:	VHS
Max running time:	180min
Recording level:	Automatic
Still picture:	Yes (single frame advance)
Slow motion:	No
Tuner:	Yes
Timer:	7 days/5 programmes
Audio dub:	Yes
Electronic edit:	nia
Tape counter:	3 digit
Inputs:	Aerial; video; audio (mic and line)
Outputs:	Aerial; video; audio (line)
Power supply:	240V
Weight:	14kg
Dimensions:	478 x 158 x 359mm
Cost:	nia
Accessories:	None
Other models:	VT-5000ER – as above with 10 day timer
Notes:	Remote pause control

Make:	**Hitachi**
Model:	VT-7000E
Tape:	VHS
Max running time:	180min
Recording level:	Automatic
Still picture:	Yes (single frame advance)
Slow motion:	No
Tuner:	No
Timer:	No
Audio dub:	Yes
Electronic edit:	Assemble
Tape counter:	3 digit
Inputs:	Video; audio (mic and line)
Outputs:	Video; audio (line)
Power supply:	12V DC
Weight:	5.9kg
Dimensions:	269 x 120 x 318mm
Cost:	(g)
Accessories:	AC adaptor; battery pack; tuner/timer
Notes:	Remote control for all functions

Make:	**Hitachi**
Model:	VT 8000E
Tape:	VHS
Max running time:	180min
Recording level:	Automatic
Still picture:	Yes (single frame advance)
Slow motion:	No
Tuner:	8 channel
Timer:	10 day/1 programme
Audio dub:	Yes
Electronic edit:	No
Tape counter:	3 digit
Inputs:	Aerial; video; audio (mic and line)
Outputs:	Aerial; video; audio (line)
Power supply:	240V
Weight:	11kg

Dimensions: 435 x 145 x 332mm
Cost: (g)
Accessories: Remote control unit
Other models: VT 8500E – as above with slow motion; fast search; wireless remote control; 7 day/5 programme timer (h)

Make: **JVC**
Model: CR-4400E
Tape: U
Max running time: 30min
Recording level: Automatic
Still picture: Yes
Slow motion: No
Tuner: No
Timer: No
Audio dub: Yes (channel 1)
Electronic edit: Automatic assemble edit
Tape counter: 3 digit
Inputs: Video-UHF; line-phono; mic-minijack
Outputs: Video-UHF; line-phono
Power supply: 12V DC
Weight: 11.2kg
Dimensions: 369 x 336 x 141mm
Cost: nia
Accessories: Battery pack; shoulder strap; crystal earphone; AC power pack adaptor
Other models: CR-4400LE – as above with BNC connectors for video and XLR connectors for audio
Notes: Optional RF convertor for CR-4400E

Make: **JVC**
Model: CR 6060E (PAL plus CCIR playback and recording standards)
Tape: U
Max running time: nia
Recording level: Auto for video; auto/manual for audio
Still picture: Yes
Slow motion: No (single frame advance)
Tuner: No
Timer: No
Audio dub: Yes (channel 1)
Electronic edit: No
Tape counter: 3 digit
Inputs: Video-UHF; audio-DIN; mic-standard jack
Outputs: Video-UHF; audio-DIN
Power supply: 110/127/220/240V
Weight: 27kg
Dimensions: 526 x 195 x 450mm
Cost: nia
Accessories: Remote control; tuner/timer; RF converter; dust cover; aerial cable; monitor cable
Other models: CR 6060ED – as above with PAL and CCIR recording; PAL, NTSC, CCIR and EIA playback standards; CR 6060ET – as above with PAL, SECAM and CCIR recording; PAL, SECAM, CCIR, NTSC and EIA playback standards
Notes: Auto search; auto replay; stereo audio

Make: **JVC**
Model: CR 8300E
Tape: U
Max running time: 60min
Recording level: Auto for video; auto/manual for audio
Still picture: Yes
Slow motion: Yes
Tuner: No
Timer: No
Audio dub: Yes
Electronic edit: Assemble and insert
Tape counter: 3 digit
Inputs: Video-UHF; audio-DIN; line-phono
Outputs: Video-UHF; audio-DIN; line-phono
Power supply: 110/127/220/240V
Weight: 31kg
Dimensions: 610 x 210 x 450mm
Cost: nia
Notes: Search facility

Make: **JVC**
Model: CR 8500 LE
Tape: U
Max running time: nia
Recording level: Automatic/manual
Still picture: Yes
Slow motion: Yes
Tuner: No
Timer: No
Audio dub: Yes
Electronic edit: Assemble and insert
Tape counter: 3 digit
Inputs: Video-BNC; audio-XLR; mic-XLR line-XLR
Outputs: Video-BNC; audio-XLR; line-XLR
Power supply: 110/127/220/240V
Weight: 41kg
Dimensions: 622 x 250 x 486mm
Cost: nia
Accessories: Automatic edit control unit (RM-85E); monitor cable; dust cover

Make: **JVC**
Model: HR 2200
Tape: VHS
Max running time: 180min
Recording level: Automatic
Still picture: Yes
Slow motion: No
Tuner: No
Timer: No
Audio dub: Yes
Electronic edit: nia
Tape counter: 4 digit
Inputs: Video; audio (mic and line)
Outputs: Video; audio (line and phone)
Power supply: 12V DC
Weight: 5.2kg (incl battery pack)
Dimensions: 288 x 103 x 267mm
Cost: nia
Accessories: Remote control unit (provided); battery pack mains adaptor; tuner/adaptor
Notes: High speed search

Make: **JVC**
Model: HR 3300 EK
Tape: VHS
Max running time: 180min
Recording level: Automatic
Still picture: No
Slow motion: No
Tuner: 8 channels
Timer: 8 day/1 programme
Audio dub: Yes
Electronic edit: No
Tape counter: 3 digit
Inputs: Aerial; video; audio (mic and line)
Outputs: Aerial; video; audio (line)
Power supply: 110/127/220/240V
Weight: 14kg
Dimensions: 453 x 147 x 337mm
Cost: nia
Accessories: None

Other models:	HR 3300 EG – as above but also tunes to VHF channels 2.12
Notes:	Pause control

Make:	**JVC**
Model:	HR 3320 EK
Tape:	VHS
Max running time:	180min
Recording level:	Automatic
Still picture:	No
Slow motion:	No
Tuner:	8 channels
Timer:	8 day/1 programme
Audio dub:	Yes
Electronic edit:	No
Tape counter:	3 digit
Inputs:	Aerial; video; audio (mic and line)
Outputs:	Aerial; video; audio (line)
Power supply:	110/127/220/240V
Weight:	14kg
Dimensions:	453 x 147 x 337mm
Cost:	nia
Accessories:	Remote control; power cord; aerial cable
Notes:	Pause control

Make:	**JVC**
Model:	HR 3660 EK
Tape:	VHS
Max running time:	180min
Recording level:	Automatic
Still picture:	Yes
Slow motion:	Yes
Tuner:	Yes (UHF channels 21-69)
Timer:	Yes
Audio dub:	Yes
Electronic edit:	No
Tape counter:	3 digit
Inputs:	Video; audio RF
Outputs:	Video; audio RF
Power supply:	110/127/220/240V
Weight:	14kg
Dimensions:	453 x 147 x 352mm
Cost:	nia
Accessories:	Power cord; dust cover; aerial cable; remote control unit
Notes:	Search facility

Make:	**JVC**
Model:	HR-4100
Tape:	VHS
Max running time:	180min
Recording level:	Automatic
Still picture:	No
Slow motion:	No
Tuner:	No
Timer:	No
Audio dub:	Yes
Electronic edit:	Assemble edit
Tape counter:	3 digit
Inputs:	Video; audio (mic and line)
Outputs:	Video; audio (line)
Power supply:	12V DC
Weight:	7.5kg
Dimensions:	338 x 137 x 328mm
Cost:	nia
Accessories:	RF converter; car battery cord; tuner/timer

Make:	**JVC**
Model:	HR-7700 EK
Tape:	VHS
Max running time:	180min
Recording level:	Automatic
Still picture:	Yes (with frame advance)
Slow motion:	Yes
Tuner:	16 channels
Timer:	2 weeks/8 programmes
Audio dub:	Yes
Electronic edit:	No
Tape counter:	4 digit
Inputs:	Aerial; video; audio (mic and line)
Outputs:	Aerial; video; audio (line)
Power supply:	110/127/220/240V
Weight:	16.5kg
Dimensions:	470 x 153 x 385mm
Cost:	nia
Accessories:	Remote control unit; mains cable; aerial cable
Notes:	High speed search facility; tape memory; auto rewind

Make:	**Mitsubishi**
Model:	HS-300B
Tape:	VHS
Max running time:	180min
Recording level:	Automatic
Still picture:	Yes
Slow motion:	Yes
Tuner:	8 channel
Timer:	7 day/6 programme
Audio dub:	Yes
Electronic edit:	No
Tape counter:	3 digit
Inputs:	Video; aerial; audio (mic and line)
Outputs:	Video; aerial; audio (line)
Power supply:	100/120/200/240V
Weight:	15.3kg
Dimensions:	498 x 162 x 346mm
Cost:	(g)
Accessories:	Infrared remote control

Make:	**National Panasonic**
Model:	NV-7000 B
Tape:	VHS
Max running time:	240min
Recording level:	Automatic
Still picture:	Yes
Slow motion:	No
Tuner:	12 channel
Timer:	14 day/8 programme
Audio dub:	Yes
Electronic edit:	No
Tape counter:	4 digit
Inputs:	Aerial; video; audio (mic and line)
Outputs:	Aerial; video; audio (line)
Power supply:	110/120/220/240V
Weight:	12.7kg
Dimensions:	480 x 136 x 356mm
Cost:	(g)
Accessories:	Infrared remote control
Notes:	Auto rewind; cue and review for fast search; cable remote control

Make:	**National Panasonic**
Model:	NV-8170 B
Tape:	VHS (PAL or NTSC standards)
Max running time:	240min
Recording level:	See note
Still picture:	Yes (single frame advance)
Slow motion:	No
Tuner:	No
Timer:	No
Audio dub:	Yes
Electronic edit:	No
Tape counter:	Yes
Inputs:	None (see note)

Outputs:	Video line (BNC 8 pin); audio line (8 pin and phonojack) audio monitor (phonojack)
Power supply:	110/120/220/240V
Weight:	15kg
Dimensions:	382 x 382 x 156mm
Cost:	(h)
Accessories:	Auto-search control; remote control; editing interface
Notes:	This is a replay machine only

Make:	**National Panasonic**
Model:	NV-8200 B
Tape:	VHS
Max running time:	180min
Recording level:	Video/auto; audio/metered
Still picture:	Yes (single frame advance)
Slow motion:	Yes
Tuner:	No
Timer:	No
Audio dub:	Yes
Electronic edit:	No
Tape counter:	3 digit
Inputs:	Video; audio (mic and line)
Outputs:	Video; audio (line); audio monitor
Power supply:	100-110/115-127/200-220/230-250V
Weight:	15.5kg
Dimensions:	382 x 382 x 156mm
Cost:	(i)
Accessories:	Auto search control; remote control; editing interface

Make:	**National Panasonic**
Model:	NV-8400 B
Tape:	VHS
Max running time:	180min
Recording level:	Automatic
Still picture:	Yes
Slow motion:	No
Tuner:	No
Timer:	No
Audio dub:	Yes
Electronic edit:	No
Tape counter:	3 digit
Inputs:	Video; audio (mic and line); camera
Outputs:	Video; audio (line); RF
Power supply:	12V DC with battery pack or car battery; 110/127/220/240V with AC adaptor
Weight:	9kg
Dimensions:	332 x 354 x 140mm
Cost:	(h)
Accessories:	Car/boat battery cord; rechargeable battery pack; tuner/timer

Make:	**National Panasonic**
Model:	NV-9200 B
Tape:	U (PAL, SECAM and NTSC)
Max running time:	60min
Recording level:	Automatic
Still picture:	No
Slow motion:	No
Tuner:	No
Timer:	No
Audio dub:	Yes
Electronic edit:	No
Tape counter:	3 digit
Inputs:	Video; audio (mic and line); subcarrier; sync
Outputs:	Video; audio (monitor and line); FM
Power supply:	110/127/220/240V
Weight:	34kg
Dimensions:	621 x 238 x 472mm
Cost:	(k)
Accessories:	None
Notes:	Remote control

Make:	**National Panasonic**
Model:	NV-9210 B
Tape:	U (PAL and SECAM)
Max running time:	60min
Recording level:	Automatic
Still picture:	Yes
Slow motion:	No
Tuner:	No
Timer:	No
Audio dub:	Yes
Electronic edit:	No
Tape counter:	3 digit
Inputs:	Video; audio (mic and line); subcarrier; sync
Outputs:	Video; audio (line and monitor); FM
Power supply:	110/127/220/240V
Weight:	30kg
Dimensions:	540 x 238 x 472mm
Cost:	(k)
Accessories:	Remote control
Notes:	Automatic rewind; auto search; RF converter; clock timer

Make:	**National Panasonic**
Model:	NV-9240 B
Tape:	U (PAL and SECAM)
Max running time:	60min
Recording level:	Automatic or manual
Still picture:	No
Slow motion:	No
Tuner:	No
Timer:	No
Audio dub:	Yes
Electronic edit:	No
Tape counter:	Yes
Inputs:	Video (TV, line, dub); audio (TV, line, mic); subcarrier; sync; time code
Outputs:	Video (TV, line, dub); audio (TV, line, monitor); time code; RF
Power supply:	110/120/220/240V
Weight:	35kg
Dimensions:	661 x 238 x 475mm
Cost:	(k)
Accessories:	Remote control; editing control; auto search control

Make:	**National Panasonic**
Model:	NV-9400 B
Tape:	U
Max running time:	20min
Recording level:	Automatic
Still picture:	Yes
Slow motion:	No
Tuner:	No
Timer:	No
Audio dub:	Yes
Electronic edit:	No
Tape counter:	3 digit
Inputs:	Video; audio (mic and line)
Outputs:	Video; audio (line)
Power supply:	12V DC (internal rechargeable battery)
Weight:	11.5kg
Dimensions:	336 x 141 x 369mm
Cost:	(k)
Accessories:	AC adaptor; car/boat cable; RF converter

Make:	**National Panasonic**
Model:	NV-9500 B
Tape:	U (PAL and SECAM)
Max running time:	60min

Recording level: Automatic or manual
Still picture: Yes
Slow motion: No
Tuner: No
Timer: No
Audio dub: Yes
Electronic edit: Insert and assemble
Tape counter: 3 digit
Inputs: Video (line, TV); audio (line, mic, TV); subcarrier; sync
Outputs: Video; audio (line, TV, monitor); FM
Power supply: 110/127/220/240V
Weight: 36kg
Dimensions: 645 x 238 x 472mm
Cost: (k)
Accessories: Remote control; edit control

Make: **National Panasonic**
Model: NV-9600 B
Tape: U
Max running time: 60min
Recording level: Automatic or manual
Still picture: Yes
Slow motion: No
Tuner: No
Timer: No
Audio dub: Yes
Electronic edit: Insert and assemble
Tape counter: Electronic digital
Inputs: Video (TV, line, dub); audio (TV, line, mic); subcarrier; sync; time code
Outputs: Video (TV, line, dub); audio (TV, line, monitor); time code; RF
Power supply: 110/120/220/240V
Weight: 35kg
Dimensions: 661 x 475 x 238mm
Cost: (k)
Accessories: Remote control; edit control; auto search control

Make: **Nordemende**
Model: V 200
Tape: VHS
Max running time: 180min
Recording level: Automatic
Still picture: Yes
Slow motion: Yes
Tuner: 12 channel
Timer: 8 day/1 programme
Audio dub: Yes
Electronic edit: No
Tape counter: 3 digit
Inputs: Video; aerial; audio (mic and line)
Outputs: Video; aerial
Power supply: 110/127/220/240V
Weight: 13.9kg
Dimensions: 453 x 147 x 352mm
Cost: nia
Accessories: None

Make: **Nordemende**
Model: V 250
Tape: VHS
Max running time: 180min
Recording level: Automatic
Still picture: No
Slow motion: No
Tuner: No
Timer: No
Audio dub: Yes
Electronic edit: No
Tape counter: 3 digit
Inputs: Video; camera; audio (mic and line)
Outputs: Video; audio (line)
Power supply: 12V accumulator built-in
Weight: 7.5kg
Dimensions: 338 x 137 x 328mm
Cost: nia
Accessories: AC adaptor; timer/tuner

Make: **Nordemende**
Model: VHS
Tape: VHS
Max running time: 180min
Recording level: Automatic
Still picture: No
Slow motion: No
Tuner: Yes
Timer: Yes
Audio dub: Yes
Electronic edit: No
Tape counter: 3 digit
Inputs: Video; aerial; audio (mic and line)
Outputs: Video; aerial
Power supply: 110/127/220/240V
Weight: 13.9kg
Dimensions: 453 x 147 x 337mm
Cost: nia
Accessories: None

Make: **Philips**
Model: N1502
Tape: VCR
Max running time: 60min
Recording level: Automatic
Still picture: Yes
Slow motion: No
Tuner: 8 channel
Timer: 3 day/1 programme
Audio dub: Yes
Electronic edit: No
Tape counter: 3 digit
Inputs: Aerial; audio
Outputs: Aerial; audio
Power supply: 220/240V
Weight: 16kg
Dimensions: 560 x 370 x 160mm
Cost: (g)
Accessories: None
Other models: N 1512 – as above with video input/output (h)

Make: **Philips**
Model: VR 20 20
Tape: Compact video cassette
Max running time: 2 x 240min
Recording level: Automatic
Still picture: No
Slow motion: No
Tuner: 26 channels
Timer: 16 day/5 programme
Audio dub: No
Electronic edit: No
Tape counter: 4 digit
Inputs: nia
Outputs: nia
Power supply: 110/127/220/240V
Weight: 17.5kg
Dimensions: 540 x 365 x 152mm
Cost: (h)
Accessories: Infrared remote control

Make: **Sanyo**
Model: VTC 5500P
Tape: Betamax
Max running time: 195min

Recording level: Automatic
Still picture: Yes
Slow motion: No
Tuner: 8 channel
Timer: 7 day/5 programme
Audio dub: Yes
Electronic edit: Assemble
Tape counter: No
Inputs: Video; aerial; audio (mic and line)
Outputs: Video; aerial; audio (line)
Power supply: 240V
Weight: 15.3kg
Dimensions: 454 x 161 x 387mm
Cost: nia
Accessories: Remote control

Make: **Sanyo**
Model: VTC 9300P
Tape: Betamax
Max running time: 195min
Recording level: Automatic
Still picture: Yes
Slow motion: No
Tuner: 8 channel
Timer: 3 day/1 programme
Audio dub: Yes
Electronic edit: Assemble
Tape counter: No
Inputs: Video; aerial; audio (mic and line)
Outputs: Video; aerial; audio (line)
Power supply: 240V
Weight: 19.2kg
Dimensions: 504 x 195 x 400mm
Cost: (e)
Accessories: None

Make: **Sharp**
Model: VC-6300H
Tape: VHS
Max running time: 180min
Recording level: Automatic
Still picture: Yes
Slow motion: Yes
Tuner: 12 channel
Timer: 7 day/7 programme
Audio dub: Yes
Electronic edit: No
Tape counter: 4 digit
Inputs: Video; aerial; audio (mic and line)
Outputs: Video; aerial; audio (line)
Power supply: 240V
Weight: 18kg
Dimensions: 485 x 167 x 403mm
Cost: nia
Accessories: None
Notes: Remote control

Make: **Sharp**
Model: VC-7300H
Tape: VHS
Max running time: 180min
Recording level: Automatic
Still picture: No
Slow motion: No
Tuner: 12 channel
Timer: 24 hour/1 programme
Audio dub: Yes
Electronic edit: No
Tape counter: 4 digit
Inputs: Video; aerial; audio (mic and line)
Outputs: Video; aerial; audio (line)
Power supply: 240V
Weight: 14kg
Dimensions: 480 x 165 x 385mm
Cost: nia
Accessories: Remote control

Make: **Sony**
Model: BVU-50P
Tape: U
Max running time: 20min
Recording level: Video/auto; audio/manual or auto
Still picture: No
Slow motion: No
Tuner: No
Timer: No
Audio dub: No
Electronic edit: Assemble
Tape counter: Elapsed time display
Inputs: Video; camera; audio (mic); time code
Outputs: Audio (monitor)
Power supply: 12V DC
Weight: nia
Dimensions: 270 x 125 x 335mm
Cost: nia
Accessories: Battery pack; battery charger; AC adaptor; time code reader/generator

Make: **Sony**
Model: BVU-110P
Tape: U
Max running time: 20min
Recording level: Video/auto; audio/manual or audio
Still picture: No
Slow motion: No
Tuner: No
Timer: No
Audio dub: Yes
Electronic edit: Assemble
Tape counter: Elapsed time display
Inputs: Video; camera; audio (mic and line)
Outputs: Video; audio (line and monitor); time code; RF
Power supply: 12V DC
Weight: 11.3kg
Dimensions: 336 x 135 x 393mm
Cost: nia
Accessories: Battery pack; battery charger; AC adaptor; time code generator

Make: **Sony**
Model: BVU-200P
Tape: U
Max running time: 60min
Recording level: Manual (metered)
Still picture: No
Slow motion: No
Tuner: No
Timer: No
Audio dub: Yes
Electronic edit: Assemble and insert
Tape counter: 3 digit
Inputs: Video; audio (mic and line); sync; time code
Outputs: Video; audio (monitor and line); monitor; remote; time code
Power supply: 100/120/220/240V
Weight: 46kg
Dimensions: 464 x 275 x 646mm
Cost: nia
Accessories: Remote control unit

Make: **Sony**
Model: SL C7UB
Tape: Betamax
Max running time: 195min

Recording level:	Automatic
Still picture:	Yes (single frame advance)
Slow motion:	Yes
Tuner:	12 channels
Timer:	14 day/4 programmes
Audio dub:	Yes
Electronic edit:	Assemble
Tape counter:	nia
Inputs:	Video; aerial; audio (mic and line)
Outputs:	Video; aerial; audio (line)
Power supply:	240V
Weight:	15.5kg
Dimensions:	485 x 163 x 379mm
Cost:	nia
Accessories:	None
Notes:	Auto programme search; infra-red remote control

Make:	**Sony**
Model:	SL 3000 UB
Tape:	Betamax
Max running time:	140min
Recording level:	Automatic
Still picture:	No
Slow motion:	No
Tuner:	No
Timer:	No
Audio dub:	No
Electronic edit:	No
Tape counter:	3 digit
Inputs:	Camera
Outputs:	Video; audio
Power supply:	12V DC
Weight:	9.1kg
Dimensions:	296 x 128 x 345mm
Cost:	nia
Accessories:	None

Make:	**Sony**
Model:	SL 8080 UB
Tape:	Betamax
Max running time:	195min
Recording level:	Automatic
Still picture:	Yes
Slow motion:	nia
Tuner:	8 channel
Timer:	3 day/1 programme
Audio dub:	nia
Electronic edit:	nia
Tape counter:	nia
Inputs:	Video; aerial; audio (mic and line)
Outputs:	Video; aerial; audio (line)
Power supply:	240V
Weight:	19.5kg
Dimensions:	522 x 189 x 394mm
Cost:	nia
Accessories:	None

Make:	**Sony**
Model:	VO 2630
Tape:	U
Max running time:	60min
Recording level:	Video/auto; audio/manual or auto
Still picture:	Yes
Slow motion:	No
Tuner:	No
Timer:	No
Audio dub:	Yes
Electronic edit:	No
Tape counter:	3 digit
Inputs:	Video; audio (mic and line)
Outputs:	Video; audio (monitor and line); RF
Power supply:	100/110/120/127/220/240V
Weight:	31.5kg
Dimensions:	591 x 224 x 416mm
Cost:	(k)
Accessories:	Remote control; tuner; RF modulator

Make:	**Sony**
Model:	VO 2631
Tape:	U
Max running time:	60min
Recording level:	Video/auto; audio/manual or auto
Still picture:	Yes
Slow motion:	No
Tuner:	No
Timer:	No
Audio dub:	Yes
Electronic edit:	No
Tape counter:	3 digit
Inputs:	Video; audio
Outputs:	Video; audio (line and monitor); RF
Power supply:	100/110/120/127/220/240V
Weight:	31.5kg
Dimensions:	591 x 224 x 416mm
Cost:	(k)
Accessories:	Auto search control; remote control; timer; RF modulator

Make:	**Sony**
Model:	VO 2860P
Tape:	U
Max running time:	60min
Recording level:	Automatic or manual
Still picture:	No
Slow motion:	No
Tuner:	No
Timer:	No
Audio dub:	Yes
Electronic edit:	Assemble, insert
Tape counter:	3 digit
Inputs:	Video; audio (mic and line)
Outputs:	Video; audio (monitor and line); RF
Power supply:	100/110/120/127/220/240V
Weight:	36.5kg
Dimensions:	646 x 226 x 462mm
Cost:	(k)
Accessories:	Auto edit control; timer

Make:	**Sony**
Model:	VO 4800 PS
Tape:	U
Max running time:	20min
Recording level:	Video/auto; audio/manual or auto
Still picture:	No
Slow motion:	No
Tuner:	No
Timer:	No
Audio dub:	Yes
Electronic edit:	Assemble
Tape counter:	3 digit
Inputs:	Video; audio (mic and line)
Outputs:	Video; audio (monitor and line); RF
Power supply:	12V DC or with mains adaptor supplied
Weight:	8kg
Dimensions:	336 x 135 x 354mm
Cost:	(k)
Accessories:	Battery pack

Make:	**Sony**
Model:	VP 2031
Tape:	U
Max running time:	60min
Recording level:	See note
Still picture:	Yes
Slow motion:	No

Tuner:	No
Timer:	No
Audio dub:	No
Electronic edit:	No
Tape counter:	3 digit
Inputs:	See note
Outputs:	Video; audio (line and monitor); RF
Power supply:	100/110/120/127/220/240V
Weight:	30kg
Dimensions:	591 x 224 x 416mm
Cost:	(k)
Accessories:	Auto search control; remote control; timer RF modulator
Notes:	Playback only

Make:	**Technicolor**
Model:	212
Tape:	¼in compact video cassette
Max running time:	30min
Recording level:	Automatic
Still picture:	Yes
Slow motion:	No
Tuner:	No
Timer:	No
Audio dub:	Yes
Electronic edit:	No
Tape counter:	3 digit
Inputs:	Video; (camera); audio (mic)
Outputs:	Video; audio; RF
Power supply:	240V AC or 12V DC
Weight:	3.3kg
Dimensions:	111 x 235 x 76mm
Cost:	nia
Accessories:	None

Make:	**Toshiba**
Model:	V-5470B
Tape:	Betamax
Max running time:	195min
Recording level:	Automatic
Still picture:	Yes
Slow motion:	Yes
Tuner:	10 channel
Timer:	7 day/3 programme
Audio dub:	Yes
Electronic edit:	nia
Tape counter:	4 digit
Inputs:	Video; aerial; audio (mic and line)
Outputs:	Video; aerial; audio (line)
Power supply:	240V
Weight:	13.5kg
Dimensions:	475 x 178 x 386mm
Cost:	(g)
Accessories:	None

VIDEO CAMERAS

The video cameras included below are principally intended to be used in conjunction with a video cassette recorder. Because of this, the majority are portable models, although a few are suitable for studio use. Both black and white and colour models are included.

The type of pick-up tube is indicated, the majority using vidicon tubes which are relatively cheap and can be used for well-lit scenes but have a certain amount of image retention when working in low light conditions.

The cheaper colour cameras have a single pick-up tube and use filters in front of the tube to separate the primary colours while the more expensive ones use more than one tube with associated optics to separate the colours.

Two types of viewfinder are in use: the optical finder allows the operator to 'see' through the camera lens by means of a system of prisms while the more expensive system involves a small video screen mounted on the camera which gives a black and white picture identical to the camera output. The video signal from the camera must be kept in sync and this can be achieved in two ways – internal, when the synchronising signal is generated by the camera itself, and external, when the signal has to be supplied by an outside source.

The power supply for the camera is usually obtained from the VCR, a battery or a mains adaptor since the required voltage is usually 12V.

Make:	**Akai**
Model:	VC30
Type:	Portable, colour
Tube:	17mm, vidicon with stripe filter
Sync:	Internal
Horizontal resolution:	250 lines
Sensitivity:	Min 100 lux
Viewfinder:	TTL optical
Power supply:	12V DC
Weight:	1.4kg
Dimensions:	80 x 232 x 269mm
Cost:	nia
Accessories:	None
Other models:	VC60 – as above with 6 x 200m lens

Make:	**Akai**
Model:	VC7100
Type:	Portable, colour
Tube:	2 x 17mm, vidicons
Sync:	Internal
Horizontal resolution:	400 lines at centre
Sensitivity:	Min 100 lux
Viewfinder:	38mm B/W tube
Power supply:	12V DC
Weight:	3.7kg
Dimensions:	nia
Cost:	(h)
Accessories:	None

Make:	**Amcron (MacInnes Labs Ltd)**
Model:	DC 300 A
Type:	Portable, colour
Tube:	17mm, vidicon with stripe filter
Sync:	Internal
Horizontal resolution:	250 lines
Sensitivity:	Min 100 lux
Viewfinder:	TTL optical
Power supply:	120-240V 50Hz
Weight:	7.5kg
Dimensions:	483 x 178 x 248mm
Cost:	(c)
Accessories:	None
Notes:	Stereo monaural

Make:	**Ferguson**
Model:	3V04
Type:	Portable, black and white
Tube:	17mm, vidicon
Sync:	nia
Horizontal resolution:	nia
Sensitivity:	nia

Viewfinder: TTL optical
Power supply: 12-14V AC
Weight: 1.2kg
Dimensions: 202 x 68 x 250mm
Cost: nia
Accessories: AC mains power source
Notes: Built-in mic

Make: **Ferguson**
Model: 3V17
Type: Portable, colour
Tube: 17mm, single
Sync: nia
Horizontal resolution: 250 lines
Sensitivity: nia
Viewfinder: TTL optical
Power supply: 12V DC
Weight: 1.4kg
Dimensions: 120 x 77 x 269mm
Cost: nia
Accessories: None
Notes: Colour temperature switch; built-in mic

Make: **Ferguson**
Model: 3V20
Type: Portable, colour
Tube: 17mm, vidicon with stripe filter
Sync: nia
Horizontal resolution: More than 270 lines
Sensitivity: nia
Viewfinder: 38mm B/W tube
Power supply: 12V DC
Weight: 2.2kg
Dimensions: 270 x 110 x 355mm
Cost: nia
Accessories: None
Notes: Colour temperature compensation; built-in mic

Make: **Grundig**
Model: FA 123
Type: Fixed, black and white
Tube: 17mm, vidicon
Sync: nia
Horizontal resolution: 440 lines at centre
Sensitivity: Min 20 lux
Viewfinder: nia
Power supply: 18V DC
Weight: 0.8kg (without lens)
Dimensions: 75 x 75 x 200mm (without lens)
Cost: nia
Accessories: None
Notes: Very small camera for monitoring or training purposes

Make: **Grundig**
Model: FA 1007
Type: Portable, black and white
Tube: 17mm, vidicon
Sync: nia
Horizontal resolution: nia
Sensitivity: nia
Viewfinder: Optical
Power supply: 12V DC
Weight: 1.2kg
Dimensions: 74 x 212 x 240mm
Cost: nia
Accessories: None
Notes: Zoom lens; built-in mic

Make: **Grundig**
Model: FAC 1800
Type: Portable, colour
Tube: 25mm, vidicon with stripe filter
Sync: Internal
Horizontal resolution: nia
Sensitivity: nia
Viewfinder: 38mm B/W tube
Power supply: nia
Weight: 2.5kg
Dimensions: 90 x 233 x 431mm
Cost: nia
Accessories: Remote control; camera extension cable
Notes: 6:1 zoom lens

Make: **Hitachi**
Model: FP-205
Type: Portable, colour
Tube: 3 x 17mm, saticons
Sync: Internal/external
Horizontal resolution: 500 lines at centre
Sensitivity: Min 150 lux
Viewfinder: 38mm, B/W tube
Power supply: 12V DC
Weight: 5.5kg
Dimensions: nia
Cost: nia
Accessories: None

Make: **Hitachi**
Model: FP-605
Type: Studio, colour
Tube: 3 x 25mm, saticon
Sync: Internal/external
Horizontal resolution: 600 lines at centre
Sensitivity: Min 50 lux
Viewfinder: 178mm, B/W tube
Power supply: 120/220/240V
Weight: 2.6kg
Dimensions: nia
Cost: nia
Accessories: Int sync unit; genlock unit; high-grade viewfinder; enhancer unit; masking unit

Make: **Hitachi**
Model: FP-1020
Type: Portable, colour
Tube: 3 x 17mm, saticons
Sync: Internal/external
Horizontal resolution: 500 lines at centre
Sensitivity: Min 200 lux
Viewfinder: 38mm, B/W tube
Power supply: 117V, 220/240V AC or 12V DC
Weight: 7kg
Dimensions: nia
Cost: nia
Accessories: None

Make: **Hitachi**
Model: FP-3060A
Type: Portable, colour
Tube: 25mm, saticons
Sync: Internal/external
Horizontal resolution: 270 lines
Sensitivity: 2000 lux, standard
Viewfinder: 38mm, B/W tube
Power supply: 12V DC
Weight: 5kg
Dimensions: nia
Cost: nia
Accessories: None

Make:	**Hitachi**
Model:	GP-7
Type:	Portable, colour
Tube:	17mm, tri-electrode vidicon
Sync:	Internal/external
Horizontal resolution:	250 lines
Sensitivity:	Min 100 lux
Viewfinder:	38mm, B/W tube
Power supply:	12V DC
Weight:	5.2kg
Dimensions:	nia
Cost:	nia
Accessories:	AC adaptor; operations panel OP7

Make:	**JVC**
Model:	GC-4100E
Type:	Portable, colour
Tube:	2 x 17mm, vidicons
Sync:	Internal
Horizontal resolution:	400 lines at centre
Sensitivity:	Min 100 lux
Viewfinder:	38mm, B/W tube
Power supply:	12V DC
Weight:	3.7kg
Dimensions:	190 x 240 x 395mm
Cost:	nia
Accessories:	None

Make:	**JVC**
Model:	GS-1000E
Type:	Portable, black and white
Tube:	17mm, vidicon
Sync:	Internal
Horizontal resolution:	450 lines
Sensitivity:	Min 50 lux
Viewfinder:	TTL optical
Power supply:	12V DC
Weight:	1.2kg
Dimensions:	68 x 202 x 250mm
Cost:	nia
Accessories:	AC adaptor; telecine attachment; camera stand

Make:	**JVC**
Model:	CY-8800E
Type:	Portable/studio, colour
Tube:	3 x 17mm, plumbicons or saticons
Sync:	Internal/external
Horizontal resolution:	500 lines at centre
Sensitivity:	Min 250 lux
Viewfinder:	38mm, B/W tube
Power supply:	12V DC
Weight:	6.3kg
Dimensions:	145 x 195 x 465mm
Cost:	nia
Accessories:	Remote control unit

Make:	**National Panasonic**
Model:	WV-3300E
Type:	Portable, colour
Tube:	25mm, vidicon with stripe filter
Sync:	Internal
Horizontal resolution:	250 lines at centre
Sensitivity:	Min 100 lux
Viewfinder:	38mm, B/W tube
Power supply:	12V DC
Weight:	2.5kg
Dimensions:	90 x 430 x 233mm
Cost:	nia
Accessories:	None
Other models:	WV-3310E – as above but with optical viewfinder

Make:	**Nordemende**
Model:	C220
Type:	Portable, colour
Tube:	17mm, vidicon with stripe filter
Sync:	Internal
Horizontal resolution:	230 lines
Sensitivity:	nia
Viewfinder:	TTL optical
Power supply:	12V DC
Weight:	1.7kg
Dimensions:	nia
Cost:	nia
Accessories:	AC adaptor; tripod; microphone; floodlight; slide to video adaptor

Make:	**Sanyo**
Model:	VCC 545P
Type:	Portable, colour
Tube:	25mm, tri-electrode vidicon
Sync:	Internal
Horizontal resolution:	250 lines at centre
Sensitivity:	Min 100 lux
Viewfinder:	Electronic
Power supply:	12V DC
Weight:	3.5kg
Dimensions:	200 x 284 x 416mm
Cost:	(h)
Accessories:	Mains adaptor
Notes:	Colour temperature compensation

Make:	**Sony**
Model:	DXC 1640P
Type:	Portable, colour
Tube:	17mm, trinicon
Sync:	Internal/external
Horizontal resolution:	300 lines
Sensitivity:	Min 100 lux
Viewfinder:	38mm, B/W tube
Power supply:	12V DC
Weight:	4.9kg
Dimensions:	nia
Cost:	(k)
Accessories:	None

Make:	**Sony**
Model:	HVC 2000P
Type:	Portable, colour
Tube:	17mm, trinicon
Sync:	Internal
Horizontal resolution:	300 lines
Sensitivity:	Min 70 lux
Viewfinder:	38mm, B/W tube
Power supply:	nia
Weight:	2.7kg
Dimensions:	225 x 200 x 351mm
Cost:	nia
Accessories:	AC adaptor; microphone; ext cable; tripod
Other models:	HVC 2010P – similar to above with optical viewfinder

SCREENS

There are two main types of projection screen:

1. *Front projection screen.* This type is the most usual and consists of a reflecting surface on to which the light from the projector is beamed. The image is viewed by the reflected light from the surface. There are several different types of surface. The matt surface reflects the least amount of light but is viewable from any position in a wide arc without much variation in brightness. The silver surface reflects more light but has a smaller viewing arc while the beaded surface reflects even more light but with an even more restricted viewing arc.
2. *Back projection screen.* In this case the screen is placed between the audience and the projector and consists of a translucent material, either plastic or frosted glass. The image is viewed by the light transmitted through the screen. As a result the image is extremely bright when viewed from within a narrow angle in front of the screen. This makes back projection screens very useful in conditions with high ambient light. Some back projection screens are fitted with a mirror which reduces the space required behind the screen by making it possible to place the projector at one side and at the same time reverses the image to compensate for the reversal produced by the projector/screen system.

Most manufacturers produce many screens with a large variety of sizes and variations in the type of mounting, free standing (tripod screens), wall mounting, ceiling mounting and electrically operated screens. Because of this, in the list which follows only the extremes of the range produced and the price range is indicated in order to make the listing manageable.

FRONT PROJECTION SCREENS

Make: **E J Arnold**
Screen sizes: 102 x 102cm to 183 x 183cm (4)
Surface: nia
Mounting: Free; tripod
Cost: (a)

Make: **Audio Visual Equipment Ltd**
Screen sizes: 112 x 173cm to 356 x 356cm (18)
Surface: Matt
Mounting: Free standing
Cost: (b) to (f)

Make: **Bell and Howell, Filmo screens**
Screen sizes: 145 x 145cm to 180 x 240cm (18)
Surface: Matt, cleanable
Mounting: Tripod; wall/ceiling; electric
Cost: (a) to (i)

Make: **Cine Screens Ltd**
Screen sizes: 100 x 100cm to 360 x 360cm (61)
Surface: Matt; pearl; lenticular
Mounting: Tripod; wall/ceiling; electric
Cost: (a) to 'price on application'

Make: **Elf, Cee; screens**
Screen sizes: 127 x 127cm to 366 x 366cm (60)
Surface: Matt; beaded; lenticular
Mounting: Tripod; wall/ceiling; electrically operated
Cost: (a) to (g)

Make: **Elite**
Screen sizes: 125 x 125cm to 240 x 180cm (10)
Surface: Matt
Mounting: Tripod; wall/ceiling
Cost: (a)

Make: **Harkness**
Screen sizes: 127 x 127cm to 426 x 320cm (24)
Surface: Silver; matt
Mounting: Tripod; wall/ceiling
Cost: (a) to (e)

Make: **ITM, Oray screens**
Screen sizes: 125 x 125cm to 300 x 400cm

Surface: Matt
Mounting: Tripod; wall/ceiling; electric
Cost: (a) to (b)

Make: **MW**
Screen sizes: 100 x 100cm to 304 x 304cm (137)
Surface: Matt; beaded; silver; pearl
Mounting: Tripod; wall/ceiling; electric
Cost: (a) to (f)

Make: **Nobo**
Screen sizes: 120 x 120cm to 177 x 177cm (5)
Surface: Lustre
Mounting: Tripod; wall
Cost: (a)

Make: **The Perforated Front Projection Co Ltd**
Screen sizes: Made to company specifications
Surface: Made to company specifications
Mounting: Made to company specifications
Cost: nia

Make: **Photax**
Screen sizes: 120 x 120cm to 177 x 177cm (8)
Surface: nia
Mounting: Tripod; wall/ceiling
Cost: (a)

Make: **Projection and Display Services Ltd**
Screen sizes: 137 x 137cm to 3657 x 3657cm (16)
Surface: Matt
Mounting: Free standing
Cost: (b) to (e)

Make: **Rank Aldis**
Screen sizes: 127 x 127cm to 177 x 177cm (3)
Surface: Matt; lenticular
Mounting: Tripod; wall/ceiling
Cost: nia

Make: **Vista**
Screen sizes: 127 x 127cm to 488 x 366cm (31)
Surface: Matt; lenticular
Mounting: Tripod; wall/ceiling
Cost: (a)

Make: **Weyel**
Screen sizes: 130 x 130cm to 180 x 180cm (7)
Surface: Matt; magnetic/write-on
Mounting: Wall/free standing
Cost: (b) to (e)

Make: **Widescreen**
Screen sizes: 203 x 76cm to 270 x 102cm (9)
Surface: Hi-flex
Mounting: Tripod; wall/ceiling
Cost: (a)

BACK PROJECTION SCREENS

Make: **E J Arnold, Group screen (1)**
Screen sizes: 375 x 375mm
Mirror: Yes
Cost: (a)

Make: **Camera Talks Ltd**
Screen sizes: 25 x 17.5cm to 61 x 45.5cm (3)
Mirror: Yes
Cost: (a)

Make: **Chatsworth**
Screen sizes: 44 x 33cm to 61 x 45cm (2)
Mirror: Yes
Cost: (a)

Make: **Cine Screens Ltd**
Screen sizes: 100 x 100cm to 150 x 150cm (3)
Mirror: No
Cost: (a)

Make: **Harkness**
Screen sizes: 183 x 183cm to 426 x 320cm (8)
Mirror: No
Cost: (c) to (e)

Make: **The Perforated Front Projection Co Ltd**
Screen sizes: Made to company specifications
Mirror: Made to company specifications
Cost: nia

Make: **Premier, Groupshow**
Screen sizes: 30 x 40cm to 64 x 64cm (3)
Mirror: Yes
Cost: (a) to (b)

Make: **Projection and Display Services Ltd**
Screen sizes: 137 x 137cm to 3657 x 3657cm (16)
Mirror: No
Cost: (c) to (f)

Make: **THD**
Screen sizes: 43 x 30cm (1)
Mirror: Yes
Cost: (a)

NON-PROJECTIONAL AIDS

VISUAL PLANNING AND DISPLAY SYSTEMS

Allen Bros (Visual Aids) Ltd
4 High Street, Horley, Surrey
Tel: 02934 2631/2

Pinboards, magnetic boards, plastigraphs, display materials, flannelgraphs (hook and loop system). Alphachart – charting and planning boards (Pegboard system)

E J Arnold & Son Ltd
Butterley Street, Leeds LS10 1AX
Tel: 0532 412944

Roll up flannelgraph, felts, flock paper, magnetic rubber products, pegboard systems

Chartplan Ltd
Howdale Road, Downham Market, Norfolk PE 38 9AL
Tel: 036 63 2511

Vacuum action charts, magnetic charts

Elite Manufacturing Co Ltd
Elite Works, Station Road, Manningtree, Essex CO11 1D2
Tel: 020-639 2171

Magnetic planning systems (special clip-in system)

ESA Creative Learning Ltd
Pinnacles, PO Box 22, Harlow, Essex CMA 5AY
Tel: 0279 21131

Magnetic boards and accessories, pinboards

Graphics Industrial Art Ltd
51 Lisson Grove, London NW1 6US
Tel: 01-723 3231

Display systems

Thomas Hope Ltd
St Philips Drive, Royton, Oldham, Lancs
Tel: 061 633 3935

Magnetic boards and accessories, pinboards

Mabco
23 Delves Avenue, Tunbridge Wells, Kent
Tel: 0892 35628

Planning boards

Magiboards Ltd
42 Wates Way, Willow Lane Industrial Estate, Mitcham, Surrey CR4 4TA
Tel: 01-640 9311

Magnetic charting systems, magnetic boards and accessories, flipcharts

Magnetic Displays Ltd
91 Endcliffe Vale Road, Sheffield S10 3ET
Tel: 0742 664610

Magnetic rubber display systems and accessories

Marker Board Suppliers Ltd
Rockfield Road, Hereford HR1 2UA
Tel: 0432 6062

Magnetic timetable planning boards, magnetic boards and accessories

Modular Designs
30-32 Devonshire Road, London SE23 3ST
Tel: 01-699 4871

Flipcharts

Morol Ltd
Northumberland House, Gresham Road, Staines, Middx
Tel: 81 51958

PVC sheeting

Philip & Tacey Ltd
Andover, Hants SP10 5BR
Tel: 0264 61171

Magna-Cel magnetic display board, Cellograph display panels materials and kits, magnetic symbols

Sasco Ltd
27 Hastings Road, Bromley, Kent BR2 8NA
Tel: 01-462 2241

Visual planning systems (self-adhesive and magnetic systems)

Signal Business Systems
Pewsey, Wilts SN9 5HL
Tel: 06726 3333

Visual control boards (tailor made), special clip-in system

Unichart Ltd
15 Aldershot Road, Aldershot, Hants
Tel: 0252 28512

Control boards (T-chart card system)

Visual Aids Centre
78 High Holborn, London WC1V 6NB
Tel: 01-242 6631

Magnetic and white boards plus accessories. Flip chart easels. Flannelgraphs and and plastigraphs.

Waves
Corscombe, Nr Dorchester, Dorset DT2 0NU
Tel: 093589 248

Procon boards (magnetic system), dry white board

Wondersigns Movitex Ltd
Green Street, Enfield, Middlesex
Tel: 01-804 8251

Noticeboards and magnetic charts (changeable clip-in and pintype lettering), interchangeable timetable, Adapta-charts (plug-in system)

Woodcon Products
Peter Road, Commerce Way, Lancing, Sussex BN15 8TH
Tel: 09063 65946

Pocket panels

WRITING BOARDS

Conté UK Ltd
Park Farm Road, Folkstone, Kent
Tel: 0303 58664

Adhesive whiteboard surface (dry wipe)

E J Arnold & Son Ltd
Butterley Street, Leeds LS10 1AX
Tel: 0532 442944

Blackboards and whiteboards

Audio Visual Material Ltd
AVM House, 1 Alexandra Road, Farnborough, Hants GU14 6BU
Tel: 0252 40721

All types of wall boards

Hulton Educational Publications Ltd
Raans Road, Amersham, Bucks HP6 6JJ
Tel: 02403 4196

Blackboards with special outlines

ITM
Ashford Road, Ashford, Middlesex TW15 1XF
Tel: 69 56222

Whiteboards (dry and wet wipe)

King Cole Tube Bending Co Ltd
40 Buckland Road, Pen Mill Trading Estate, Yeovil, Somerset
Tel: 0935 26141

Portable white and green writing boards

Marker Board Supplies Ltd
Rockfield Road, Hereford HR1 2UA
Tel: 0432 6062

All types of writing boards and accessories

Philip & Tacey Ltd
North Way, Andover, Hants SP10 5BA
Tel: 0264 61171

Multiplan writing and display boards (chalk and wet wipe)

Pilot (Stationers) Ltd
Pilot House, Mallow Street, London EC1
Tel: 01-253 6806

Wyteboard (dry wipe), accessories for writing boards

Sasco Ltd
27 Hastings Road, Bromley, Kent BR2 8NA
Tel: 01-462 2241

Visual planning systems and dry wipe writing boards

Selectasize
5 Pollards Hill South, Norbury, London SW16 4LW
Tel: 01-764 6386

Whiteboards (dry and wet wipe)

Swales-Sofadi Ltd
PO Box No 10, Dock Road, North Shields NE29 6DF
Tel: 089 45 79314

Adhesive blackboard fabric

Teaching Wall Systems Ltd
185 Walton Sommit Centre, Bamber Bridge, Preston, Lancs
Tel: 0772 37249

Teaching wall systems, chalk boards, wet and dry wipe boards, rollerboards

Tidmarsh & Sons
Transenna Works, 1 Laycock Street, London W1
Tel: 01-226 2261

Revolving chalkboards, roller blinds

ITM Unichart Ltd
15 Holder Road, Aldershot, Hants
Tel: 0252 28512

Chalkboards, whiteboards (dry and wet wipe)

Vamplan Services
1 Berens Road, Kensal Green, London NW10
Tel: 01-969 2281

Whiteboards, chalkboards

Waves
Corscombe, Nr Dorchester, Dorset
Tel: 093589 248

White adhesive panels, dry whiteboard

Wilson & Garden Ltd
Newton Street, Kilsyth, Glasgow G65 0JX
Tel: 0236 823291

'Unique' revolving surface writing boards (white, black or green), electronic music chalkboards

Wondersigns Movitex Ltd
Green Street, Enfield, Middlesex
Tel: 01-804 8251

Whiteboards (wet wipe)

THERMAL COPIERS

The main reason for including thermal copiers in the Handbook is their usefulness in producing transparencies for overhead projectors. The original for such transparencies must have been produced by drawing or printing, using some form of heat absorbent ink or other material such as Indian ink, pencil or printer's ink. The copiers will only accept single sheets and these are usually limited in size to approximately A4 or slightly bigger. The transparency produced can be of various types, black on clear, coloured on clear or clear lines on a coloured base, depending on the type of master acetate used.

The copiers can also produce a variety of other copies, single paper copies, masters to be used on spirit duplicators and combined spirit masters and OHP transparencies. It is also possible to use most of the copiers to protect diagrams etc on thin card or paper by laminating a plastic sheet to the surface.

Make: **Banda**
Model: Bandaflex U
Functions: Paper copies; spirit masters; OHP transparencies; DP copies; ink stencils; laminating
Dimensions of original: Max width 280mm
Power supply: 240V
Weight: 8kg
Dimensions: 414 x 118 x 273mm
Cost: (c)
Other models: Bandaflex UF – as above with special stencil control (c)
Notes: Carrier required

Make: **Bell and Howell**
Model: TM-2000
Functions: Paper copies; spirit masters; OHP transparencies; ink stencils; laminating
Dimensions of original: nia
Power supply: 240V
Weight: nia
Dimensions: nia
Cost: (d)
Notes: Carrier required

Make: **Fordigraph**
Model: Fordifax 9
Functions: Paper copies; spirit masters; OHP transparencies; ink stencils; laminating
Dimensions of original: Max width 241mm
Power supply: 110/115 or 220-240V
Weight: 14kg
Dimensions: 438 x 162 x 368mm
Cost: (c)

Make: **Gestetner**
Model: TH6
Functions: Paper copies; spirit masters; OHP transparencies; laminating
Dimensions of original: Max width 240mm
Power supply: 240V
Weight: 11kg
Dimensions: 448 x 167 x 366mm
Cost: (d)
Notes: Carrier required

Make: **ITM**
Model: 230B
Functions: Paper copies; spirit masters; OHP transparencies; ink stencils
Dimensions of original: nia
Power supply: 220-240V
Weight: 10kg
Dimensions: 485 x 140 x 240mm
Cost: (a)
Other models: 45EBU, larger version of 230B for continuous copying (d)
Notes: Carrier required

Make: **OEM (Reprographic) Ltd**
Model: Commodore II
Functions: Paper copies; spirit masters; OHP transparencies; ink stencils; laminating
Dimensions of original: nia
Power supply: 240V
Weight: nia
Dimensions: nia
Cost: (b)

Make: **Ormig**
Model: Thermograph II
Functions: Paper copies; spirit masters; OHP transparencies; laminating
Dimensions of original: nia
Power supply: 240V
Weight: nia
Dimensions: nia
Cost: nia
Other models: Thermograph III – as above but max original width 300mm (h)

Make: **Roneo Vickers**
Model: Roneotherm
Functions: Paper copies; spirit masters; OHP transparencies; ink stencils; laminating
Dimensions of original: nia
Power supply: 240V
Weight: nia

Dimensions:	nia
Cost:	(c)
Notes:	Carrier required
Make:	**3M**
Model:	45
Functions:	Paper copies; spirit masters; OHP transparencies; ink stencils; laminating
Dimensions of original:	Max width 240mm
Power supply:	220-240V
Weight:	19kg
Dimensions:	495 x 190 x 394mm
Cost:	(d)
Notes:	No carrier required

MICROCOMPUTERS

Microcomputers are the latest addition to the range of equipment available to students, teachers and lecturers and because of this it is felt that they should be included in this edition of the Handbook. While it may be argued that micros are not audio visual hardware, there are several good reasons for the following list of those which are considered to be most useful in the field of education. It is fully realised that the entries are only a small sample of this rapidly increasing area.

The heart of any microcomputer is the Central Processing Unit (CPU) which is a single microprocessor chip and its associated circuitry. The CPU needs some form of memory to retain programs and data. In most micros two forms of memory are used. Read Only Memory (ROM). The first type can only supply information to the CPU; it is not possible for the contents of the ROM to be added to or altered by the CPU. Another feature of the ROM is its ability to retain information even when power is switched off. As a result, it is often used to store the essential information for operating the micro.

The second type of memory is the Random Access Memory (RAM). In this case the CPU can both read and alter information which is contained in the memory. The size of this memory determines the amount of information which can be processed by the micro and this size is specified in thousands of bytes (k bytes or just k). The main disadvantage of the RAM is that when the power is switched off its contents are lost. To compensate for this programs and data can be saved for future use in two ways:

1. *On cassette.* In this case the information is recorded in magnetic form on standard audio cassette, often using a normal audio cassette recorder with an appropriate adaptor or interface.
2. *On floppy disc.* This is a magnetic disc made of a soft, non-rigid material – hence its name. Floppy discs can retain more information than cassettes, and the information can be read from them much more rapidly. However, they require specialised equipment to use them.

Most micros display the information on small TV type screens called Visual Display Units (VDU) although some can be connected to existing TV sets while most can also be connected to a printer which will give a permanent record of the information. All computers are given instructions in one or more of several computer languages and micros are no exception. In the listing these are given under 'software'. Those given as 'supplied' come with the micro while those listed as 'optional' must be purchased as extras.

Make and Model: **Aim 65C**
CPU: 6502
RAM: 1-4k
Cassette: 2 x interface
Floppy disc: Can be added
VDU: No
Software supplied: ASSEMBLER; TEXT EDITOR
Software optional: BASIC; PL65
Cost: (c)
Notes: 20 character LED display; 20 character thermal printer

Make and Model: **Altos ACS 8000**
CPU: Z80
RAM: 64k
Cassette: No
Floppy disc: Dual 8in
VDU: No
Software supplied: Control programme for micros
Software optional: BASIC; C BASIC; COBOL; PASCAL; FORTRAN
Cost: (k)
Notes: Expandable to a 4 user system

Make and Model: **Acorn Atom**
CPU: 6502
RAM: 2-11k
Cassette: Interface
Floppy disc: No
VDU: Interface
Software supplied: BASIC; ASSEMBLER; cassette operating system
Cost: (b)
Notes: High resolution graphics available; loudspeaker output; colour monitor output

Make and Model: **Apple II**
CPU: 6502
RAM: 16-48k
Cassette: nia
Floppy disc: Optional 5¼in expands to 116k
VDU: nia
Software supplied: Operating system; BASIC
Software optional: PASCAL; FORTRAN
Cost: (g)
Notes: High resolution graphics available

Make and Model: **Commodore Pet**
CPU: 6502
RAM: 8-32k
Cassette: Yes
Floppy disc: Dual 5¼in optional
VDU: 25 line, 40 character
Software supplied: Operating system; BASIC
Software optional: FORTH; PILOT; PASCAL
Cost: (e)
Notes: 3 models available with 8k, 16k or 32k RAM; new model with 80 character screen

Make and Model: **Compucolor II**
CPU: 8080
RAM: 8-32k
Cassette: No
Floppy disc: Single 5¼in
VDU: 32 line, 64 character, 8 colour

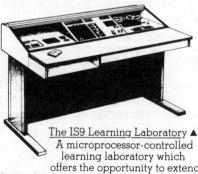

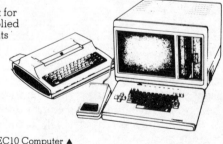

Software supplied:	Disc operating system; EX BASIC; ASSEMBLER
Cost:	(j)
Notes:	High resolution graphics available

Make and Model:	**Cromemco System 2**
CPU:	Z80
RAM:	64k
Cassette:	No
Floppy disc:	Dual 5¼in
VDU:	No
Software supplied:	Disc operating system
Software optional:	BASIC; COBOL; FORTRAN; RPG II; LISP; ASSEMBLER; multi-user BASIC
Cost:	(k)
Other models:	System 3 with dual 8in floppy disc; System Z2H with hard disc
Notes:	Expandable to multi-user (7 max); dual 8in floppy disc optional

Make and Model:	**Equinox 200**
CPU:	Z80
RAM:	64-256k
Cassette:	No
Floppy disc:	No, but hard disc available
VDU:	No
Software supplied:	Control programme for micros; C BASIC
Software optional:	COBOL; FORTRAN
Cost:	(k)

Make and Model:	**Exidy Sorcerer**
CPU:	Z80
RAM:	16-48k
Cassette:	No
Floppy disc:	Single 5¼in optional
VDU:	30 line, 64 character
Software supplied:	Operating system; BASIC
Software optional:	TEXT EDITOR; ASSEMBLER; control programme for micros; ALGOL; FORTRAN; BASIC
Cost:	(h)
Notes:	High resolution graphics available

Make and Model:	**Hewlett Packard HP 85**
CPU:	nia
RAM:	16-32k
Cassette:	No
Floppy disc:	Dual 5¼in or dual 8in optional
VDU:	16 line, 32 character
Software supplied:	BASIC
Software optional:	None
Cost:	(k)
Notes:	64 character per second printer; full dot matrix graphics; complete range of application packages, interfaces and peripherals available

Make and Model:	**Newbrain MB**
CPU:	Z80A
RAM:	2-4k
Cassette:	Interface
Floppy disc:	No
VDU:	14 line, 16 character
Software supplied:	C BASIC
Software optional:	None
Cost:	(c)
Notes:	Graphics available; battery or mains version

Make and Model:	**North Star Horizon**
CPU:	Z80A
RAM:	48-56k
Cassette:	No
Floppy disc:	Dual 5¼in
VDU:	24 line, 80 character
Software supplied:	DISC OPERATING SYSTEM; BASIC; control programme for micros
Software optional:	COBOL; FORTRAN; PASCAL
Cost:	(k)

Make and Model:	**Rair Black Box**
CPU:	8085
RAM:	32-64k
Cassette:	No
Floppy disc:	Dual 5¼in
VDU:	No
Software supplied:	Control programme for micros; BASIC; COBOL; FORTRAN; MACRO ASSEMBLER
Software optional:	None
Cost:	(k)
Notes:	Hard disc drive available

Make and Model:	**Research Machines 380Z**
CPU:	Z80A
RAM:	16-56k
Cassette:	Yes
Floppy disc:	Dual 8in optional
VDU:	No
Software supplied:	EX BASIC; ASSEMBLER; TEXT EDITOR; UTILITY
Software optional:	Control programme for micros; FORTRAN; COBOL; ALGOL; CECIL
Cost:	(k)
Notes:	Many possible systems can be built up

Make and Model:	**Sharp MZ-80k**
CPU:	Z80
RAM:	6-34k
Cassette:	Yes
Floppy disc:	Dual 5¼in optional
VDU:	24 line, 40 character
Software supplied:	BASIC; ASSEMBLER
Software optional:	None
Cost:	(e)
Notes:	Graphics facility; loudspeaker

Make and Model:	**Sinclair ZX 80**
CPU:	Z80A
RAM:	1-16k
Cassette:	Interface
Floppy disc:	No
VDU:	TV interface
Software supplied:	BASIC
Software optional:	None
Cost:	(a)
Notes:	Available in kit form

Make and Model:	**Superbrain**
CPU:	2 x ZX80
RAM:	64k
Cassette:	No
Floppy disc:	Dual 5¼in; dual 8in optional
VDU:	25 line, 80 character
Software supplied:	Control programme for micros; ASSEMBLER
Software optional:	BASIC; COBOL; FORTRAN; PASCAL
Cost:	(k)
Notes:	Limited graphics

Make and Model:	**Tandberg EC10 (2115)**
CPU:	Intel 8080 series
RAM:	56-64 K bytes
Cassette:	No
Floppy disc:	One integral 8in sugart plus 3 x 8 external
VDU:	Integral 24 x 80
Software supplied:	Multi-user BASIC, ASSEMBLER
Software optional:	CP/M, COBOL, PASCAL, C BASIC, Admin Package, Cal Package
Cost:	(k)
Notes:	Tandberg EC10 supports eight users (VDU) or Teletypes

Make and Model:	**Tandy TR S80**
CPU:	Z80
RAM:	4-16k
Cassette:	Yes
Floppy disc:	No
VDU:	16 line, 64 character
Software supplied:	BASIC; ASSEMBLER
Software optional:	None
Cost:	(d)
Notes:	Many extras available

Make and Model:	**TI 99/4**
CPU:	9900
RAM:	16k
Cassette:	2 x interface
Floppy disc:	No
VDU:	24 line, 32 character
Software supplied:	Operating system BASIC
Software optional:	None
Cost:	(h)
Notes:	TV interface (colour)

MANUFACTURERS' NAMES AND ADDRESSES

Agfa-Gevaert Ltd
27 Great West Road
Brentford
Middlesex TW8 9AX
01-560 2131

Aiwa (UK) Ltd
Aiwa House
30 Concord Road
Westwood Park Trading Estate
Western Avenue
London W3 0TH
01-993 1672

E J Arnold & Son Ltd
Butterley Street
Leeds LS10 1AX
0532 442944

Audio Visual Equipment Ltd
73 Surbiton Road
Kingston
Surrey KT1 2HG
01-549 3464

Audix Ltd
Station Road
Wenden
Saffron Waldon
Essex CB11 4L9
0799 40888

Avcom Systems Ltd
Newton Works
Stanlake Mews
London W12 7HS
01-749 2201

Bauer
Photopia International
Hempstalls Lane
Newcastle
Staffordshire ST5 0SW
0782 615131

Beaulieu
Maison Brandt Freres
16 rue de la Cerisaie
94220 Charenton le Pont
France

Bell & Howell AV Ltd
Alperton House
Bridgewater Road
Wembley
Middlesex HA0 1EG
01-903 5411

Bergen Expo Systems Inc
1088 Main Avenue
Clifton
New Jersey
USA

Bolex
14 Priestley Way
London NW2 7TN
01-450 8070

Braun Electric (UK) Ltd
Dolphin Estate
Windmill Road
Sunbury on Thames
Middlesex
09327 85611

Camera Talks Ltd
31 North Row
London W1R 2EN
01-493 2761

Clear Light
Mynad AV Sales Ltd
10-11 Great Newport Street
London WC2
01-240 1941

Coomber Electronic Equipment Ltd
58 The Tything
Worcester WR1 1JT
0905 25168

Dixons Professional
7 Dean Street
London W1V 5DE
01-734 2116

Draper Screen Co
Spiceland
Indiana 47385
USA

Dukane
Cintron Group
Grove House
551 London Road
Isleworth
Middlesex TW7 4DS
01-568 0131

Eagle International
Precision Centre
Heather Park Drive
Wembley HA0 1SU
01-902 8832

Edric Audio Visual Ltd
34-36 Oak End Way
Gerrards Cross
Buckinghamshire SL9 8BR
02813 84646

Electrosonic
815 Woolwich Road
London SE7 8LT
01-855 1101

Elf Audio Visual Ltd
836 Yeovil Road
Trading Estate
Slough
Berkshire
0753 36123

Elite Optics Ltd
354 Caerphilly Road
Cardiff CF4 4XJ
0222 63201

Elmo
C Z Scientific Instruments Ltd
2 Elstree Way
Borehamwood
Hertfordshire WD6 2AR
01-953 1688

Erskine Westayr (Engineering) Ltd
PO Box 16
Irvine
Ayrshire KA12 8JL
0294 75211

Eumig (UK) Ltd
14 Priestley Way
London NW2 7TN
01-450 8070

Fairchild
Gordon Audio Visual Ltd
Symes Mews
37 Camden High Street
London NW1
01-388 7908

Farnell AV Ltd
Kenyon Street
Sheffield S1 4BD
0742 730231

Ferguson
Thorn EMI Ltd
Thorn House
Upper St Martin's Lane
London WC2H 9ED
01-836 2444

Fordigraph
Ofrex House
Stephen Street
London W1A 1EA
01-636 3686

Fortronic Ltd
Holden Way
Donibrostle Industrial Estate
Dunfermline
Fife KY11 5JW
0383 823121

R W Friedel & Co Ltd
6 Frogmore Road
Hemel Hempstead
Hertfordshire HP3 9RW
0442 54159

Fumeo
John King Films Ltd
Film House
71 East Street
Brighton
Sussex BN1 1NZ
0273 20267

Gateway Educational Media
Waverley Road
Yate
Bristol BS17 5RB
0272 773422

Gnome Photographic Products Ltd
Gnome Corner
354 Caerphilly Road
Cardiff CF4 4XJ
0222 63201

Goodsell Ltd
New England House
New England Street
Brighton
Sussex BN1 4GH
0273 605752

Grundig International Ltd
42 Newlands Park
Sydenham
London SE26 5NQ
01-659 2468

Hacker Sound Ltd
St James House
Roman Road
Blackburn
Lancashire
0254 53525

Harkness Screens Ltd
The Gate Studio
Station Road
Borehamwood
Hertfordshire WD6 1DQ
01-953 3611

Hitachi Sales (UK) Ltd
Hitachi House
Station Road
Hayes
Middlesex UB3 4DR
01-848 8787

Ilado
Gordon Audio Visual Ltd
Symes Mews
37 Camden High Street
London NW1
01-388 7908

Imatronic Ltd
Dorian House
Rose Street
Wokingham
Berkshire
0734 784903

ITM Unichart
15 Holder Road
Aldershot
Hampshire GU12 4RH
0252 28512

ITT Consumer Products Ltd
Chester Hall Lane
Basildon
Essex
0268 3040

JVC (UK) Ltd
Eldonwall Trading Estate
Staples Corner
6-8 Priestley Way
London NW2 7AF
01-450 2621

Kindermann & Co
J L Silber Ltd
Engineers Way
Wembley
Middlesex HA9 0EA
01-903 8081

La Belle Industries (Canada) Ltd
Fraser Peacock Associates Ltd
94 High Street
Wimbledon Village SW19 5EG
01-947 7551

Leitz (Instruments) Ltd
48 Park Street
Luton LU1 3HP
0582 413811

Liesegang
George Elliot & Sons Ltd
Ajax House
Hertford Road
Barking
Essex IG11 8BA
01-591 5599

Livesey Audio Visual
95 Princes Avenue
Hull
North Humberside HU5 3QR
0482 43453

Lumatic
Associated Visual Products
Prospect House
14-20 Prospect Place
Welwyn
Hertfordshire AL6 9EN
043-871 7931

Marantz Audio Ltd
193 London Road
Staines
Middlesex
81 50132

Microtecnia
Pulse Ltd
The Laboratory
Romsey Road
Cadnum
Southampton SO4 2NN
042-127 2869

Minolta (UK) Ltd
1-3 Tanners Drive
Blakelands
Milton Keynes
MK14 5EW
0908 615141

Mitsubishi Electric (UK) Ltd
Otterspool Way
Watford
Hertfordshire WD2 8LD
92 40566

MW Screens
AV Distributors Ltd
26 Park Road
Baker Street
London NW1 4SH
01-935 8161

National Panasonic (UK) Ltd
300-318 Bath Road
Slough
Berkshire SL1 6JB
0753 34522

Neal (North East Audio Ltd)
Simonside Works
South Shields
Tyne & Wear NE34 9NX
0632 566321

Nobo Visual Aids Ltd
Alder Close
Compton Industrial Estate
Eastbourne
East Sussex BN23 6QB
0323 641521

Nordemende
Postfach 448360
2800 Bremen 44

North London Cameras
177 Stoke Newington High Street
London N16 0LH
01-254 9582

O E M (Reprographic) Ltd
133-149 St Nicholas Circle
Leicester LE1 4LE
0533 50481

Ormig
Werk Oeynhausen
497 Bad Oeynhausen
Brunhildestrasse 18
West Germany

Ozalid (UK) Ltd
Langston Road
Loughton
Essex IG10 3TH
01-508 5544

Paul Plus Ltd
Photopia Ltd
Hempstalls Lane
Newcastle
Staffordshire ST5 0SW
0782 615131

The Perforated Front Projection Co Ltd
182 High Street
Coltenham
Nr Cambridge CB4 4RX
0954 50139

Philips Video Ltd
PO Box 298
City House
420-30 London Road
Croydon CR9 3QR
01-689 2166

Photax (London) Ltd
Hampden Park Industrial Estate
Eastbourne
East Sussex BN22 9BG
0323 51197

Pioneer Hi Fi (GB) Ltd
The Ridgeway
Iver
Buckinghamshire SL0 9JL
0753 652222

Prestinox
Aico (UK) Ltd
Aico House
Faraday Road
London Road Estate
Newbury
Berkshire RG13 2AD
0635 49797

Projection Optics Co Inc
Florham Park
New Jersey 07932
USA

Rank Aldis
Rank Audio Visual Ltd
PO Box 70
Great West Road
Brentford
Middlesex TW8 9HR
01-560 6574

Rediffusion Industrial Services Ltd
Rediffusion House
214 Red Lion Road
Surbiton
Surrey KT6 7RP
01-397 5133

Revox
FWO Bauch Ltd
49 Theobalds Road
Borehamwood
Hertfordshire WD6 4RZ
01-953 0091

Robert Rigby Ltd
Premier Works
4 Northington Street
London WC1N 2JH
01-405 2944

Rollei (UK) Ltd
Denington Estate
Wellingborough
Northamptonshire NN8 2RG
0933 76431

Roneo Vickers Ltd
South Street
Romford
Essex RM1 2AR
70 46000

Sanyo Marubeni (UK) Ltd
8 Greycaine Road
Greycaine Estate
Watford WD2 4UQ
92 46363

Sharp Electronics (UK) Ltd
Thorp Road
Newton Heath
Manchester M10 9BE
061-205 2333

Simda
Programmed Presentations (Hardware) Ltd
10 St Martins Court
London WC2N 4AJ
01-240 0819

Sinclair Research Ltd
6 Kings Parade
Cambridge CB2 1SN
0276 66104

Smiths Electrical Engineers
Baldwin Street
Bamber Bridge
Preston PR5 6SR
0772 35983

Sony (UK) Ltd
Pyrene House
Sunbury Cross
Sunbury-on-Thames
Middlesex TW16 7AT
09327 89581

Tandberg Ltd
Unit 1
Revie Road Industrial Estate
Elland Road
Leeds LS11 8JG
0532 774844

Tandy Corporation
Branch UK
Bilston Road
Wednesbury
West Midlands WS10 7JN
021-556 6101

Technicolor
PO Box 7
Bath Road
West Drayton
Middlesex UB7 0DB
01-759 5432

T H D Manufacturing Ltd
THD House
South Coast Road
Peacehaven
Sussex
079-14 5791

3M (UK) Ltd
3M House
PO Box 1
Bracknell
Berkshire RG12 1JU
0344 26726

T O A Electric Co Ltd
PO Box 82
Castle Street
Ongar
Essex
0277 364333

Toshiba (UK) Ltd
Toshiba House
Frimley Road
Frimley
Camberley
Surrey GU16 5JJ
0276 62222

Viewlex AV Inc
Lees Cameras (Holborn) Ltd
58 Holborn Viaduct
London EC1A 2FD
01-236 0081

Vista
D R H (Screens) Ltd
Marco Pen Building
Southbury Road
Ponders End
Middlesex
01-805 8176

Weyel
Teaching Wall Systems Ltd
185 Walton Summit Centre
Bamber Bridge
Preston
Lancashire PR5 8AJ
0772 37249

The Widescreen Centre
48 Dorset Street
London W1H 3FH
01-935 2580

LATE ENTRIES

Tape visual section – Slide/tape viewers

Make:	**Clear light**
Model:	Diamond presenter
Tape:	Compact cassette
Synchronising system:	Multiplex track 4
Output power:	20W x 2
Speakers:	2 in lids
Frequency range:	50-12000 HZ (± 3DB)
Tone controls:	Yes
Outputs:	Tape start/stop; line; sync; manual remote; speakers (2); projector and mains leads
Power supply:	120/240V
Weight:	14kg
Dimensions:	500 x 360 x 140mm
Cost:	(j)
Accessories:	Micro Diamond Encoder for programming onto an external tape recorder. Two projector cases.
Other models:	Star Presenter – for use with 3 projectors.
Notes:	Complete unit recessed into heavy duty travel case. All units for use with standard Kodak Carousel Projectors.

Tape visual section – Dissolve units

Make:	**Clear light**
Model:	Micro Diamond Dissolve
System:	Multiplex
Controls:	5 independent dissolve rates; freeze/hold; send home; on-off timer; screen 1-2-3
Hand control:	Remote with 2 sec dissolve/reverse/home
Connections:	Tape recorder in-out; remote hand control; 2 projectors; mains; voltage; mech home; tray size.
Power supply:	115/230V 50/60HZ
Dimensions:	64 x 152 x 216mm
Cost:	(c)
Other models:	Micro Diamond Memory Programmer; Micro Star Memory Programmer

MEDIA

PRODUCERS/DISTRIBUTORS OF FILMSTRIPS/SLIDES WITH RECORDINGS

Most will now supply filmstrips *or* slides according to needs *or* a filmstrip with slide mounts. In some cases open reel tapes are available instead of cassettes. This is by no means a comprehensive list. Where producers specialise in certain subjects these are stated after the address. *NB* EFL = English as a second foreign language.

E J Arnold and Son Ltd
Butterley Street
Leeds LS10 1AX
0532 442944

Audio Learning Ltd
Sarda House
183-189 Queensway
London W2
01-727 2748

(Upper school subjects)

Audio Visual Library Services
10-12 Powdrake Road
Grangemouth
Stirlingshire FK3 9UT
03244 71521

Audio Visual Productions
Hocker Hill House
Chepstow
Gwent NP6 5ER
02912 5439

BBC Radiovision
BBC Publications
35 Marylebone High Street
London W1M 4AA
01-580 5577 or 407 6961
(distribution)

BP Educational Service
PO Box 5
Wetherby
W Yorks LS23 7EH
0937 843477

Camera Talks
31 North Row
London W1R 2EN
01-493 2761

(Health and hygiene, social welfare, nursing, languages)

Careers and Occupational Information Centre (COIC)
Manpower Services Commission
Moorfoot
Sheffield S1 4QP

(Careers information)

Carwal Audio-Visual Aids Ltd
183 High Road
Benfleet
Essex
03745 53136

(Religion, history, social studies)

Centre for the Teaching of Reading
University of Reading School of Education
29 Eastern Avenue
Reading RG1 5RU
0734 62662

(The teaching of reading)

Chiltern Consortium
Wall Hall
Aldenham, Watford
Herts WD2 8AT
09276 5880

(Teacher training and higher education)

Christian Aid
PO Box 1
London SW9 8BH
01-733 5500

(Third world studies, Christian Aid activities)

Church Missionary Society
157 Waterloo Road
London SE1
01-928 8681

(Religious topics, missionary work)

Commonwealth Association of Architects
The Building Centre
26 Store Street
London WC1E 7BT
01-636 8276

(Architecture and building)

Diana Wyllie Ltd
3 Park Road
Baker Street
London NW1
01-723 7333

(Environmental studies, meteorological and architectural)

Dundee College of Education
Co-ordinator of Learning Resources
Gardyne Road
Broughty Ferry
Dundee DDY 1NY
0382 453433

(Teaching methods)

Educational Audio Visual
Mary Glasgow Publications Ltd
Brookhampton Lane
Kineton, Warcs CV35 0JB
0926 640606

Educational Foundation for Visual Aids (EFVA)
National Audio Visual Aids Library
Paxton Place
Gipsy Road
London SE27 9SR
01-670 4247

Educational Media International
25 Boileau Road
London W5
01-998 8657

(Primary and secondary school)

Educational Productions
Bradford Road
East Ardsley
Wakefield
Yorks
0924 823971

Edward Patterson Associates
68 Copers Cope Road
Beckenham
Kent BR3 1RJ
01-658 1515

European Schoolbooks Ltd
Crost Street
Cheltenham
Glos GL53 0HX
0242 45252

(Languages, European studies)

Fergus Davidson Associates (distributors only)
1 Bensham Lane
West Croydon
Surrey CR0 2RU
01-689 6824

Focal Point Audio Visual Limited
251 Copnor Road
Portsmouth PO3 5EE
0705 65249

Hobsons Press Ltd
Bateman Street
Cambridge CB2 1LZ
0223 69811

(CRAC careers education kits)

Hugh Baddeley Productions
64 Moffats Lane
Brookmans Park
Hatfield
Herts AL9 7RU
77 54046

(Educational)

The Industrial Society
Peter Runge House
3 Carlton House Terrace
London SW1
01-839 4300

(Management, industrial relations)

Jordanhill College of Education
Audio Visual Media Dept
76 Southbrae Drive
Glasgow G13 1PP
041-959 1232

(Teaching methods)

Longman/Common Ground Filmstrips
Longman House
Burnt Mill, Harlow, Essex
0279 26721

Macmillan Education
Houndmills
Basingstoke, Hants
0256 29242

(Industrial relations, social studies)

McGraw-Hill Book Co (UK) Ltd
Shoppenhangers Road
Maidenhead
Berks SL6 2QL
0628 23432

Mary Glasgow Publications Ltd
Brookhampton Lane
Kineton
Warcs CV35 0JB
0926 640606

Opsis Ltd
140A London Road
Southborough
Tunbridge Wells, Kent
0892 42708

Oxford Educational Resources Ltd
197 Botley Road
Oxford OX2 0HE
0865 49988/41474

(Biology, medical, health education)

Pattison and Co
Newgate
Sandpit Lane
St Albans, Herts
0727 55574/54563/52206

(Instrumentation, science and the environment)

Pavic Productions
Instructional Technology Unit
Sheffield City Polytechnic
36 Collegiate Crescent
Sheffield S10 2BP
0742 665274

(Teaching methods, industrial training)

Prismatron Productions
9 Gloucester Crescent
London NW1
01-485 3582

(Statistics and computing and operations research)

Rank Audio Visual
PO Box 20
Great West Road
Brentford
Middlesex TW8 9HR
01-568 9222

(Industrial and educational)

Road Transport Industry Training Board
Training Resources Information Unit
Capitol House
Empire Way
Wembley HA9 0NG
01-902 8880 ext 284

SB Modules
159-163 Great Portland Street
London W1N 5FD
01-323 1144

Scripture Union House
130 City Road
London EC1V 2NJ
01-250 1966

(Religion, social studies)

The Slide Centre
143 Chatham Road
London SW11
01-223 3457

Space Frontiers Ltd
30 Fifth Avenue
Havant, Hants PO9 2PL
0705 475313

(Astronomy, space exploration)

Students Recordings Ltd
88 Queen Street
Newton Abbot
Devon
0626 67026

Studio Two Educational
6 High Street
Barkway, Royston
Herts SG8 8EE
0763 84759

Training and Education Associates Ltd
41 Paradise Walk
London SW3 4JW
01-351 1151

(Sports coaching National Westminster Bank)

Training Films International Ltd
14 St Mary's Street
Whitchurch
Shropshire SY13 1QY
0948 2597

also
159 Great Portland Street
London W1
01-323 1144

(Industrial relations, safety, sales)

Viewtech Audio Visual Media (Distributors only)
122 Goldcrest Road
Chipping Sodbury BS17 6XF
0272 773422

Vistech
74 Brodrick Road
London SW17 7DY
01-672 1751

(Mainly secretarial subjects)

Visual Publications Ltd
The Green
Northleach
Cheltenham, Glos
04516 518

Weston Woods Studios Ltd
14 Friday Street
Henley-on-Thames, Oxon
04912 77033

(Children's stories)

A Wheaton and Co Ltd
(Pergamon Press)
Hennock Road
Exeter EX2 8RP
0392 74121

(History)

PRODUCERS/DISTRIBUTORS OF FILMSTRIPS/SLIDE SETS

Ann and Bury Peerless
22 King's Avenue
Minnis Bay
Birchington, Kent
0843 41428

(Environmental studies, geography, religion)

The Bodleian Library
Western Manuscript Dept
Oxford OX1 3BG
0865 44675

(Filmstrips/individual slides of the Bodleian's collection of manuscripts)

British Transport Films
Melbury House
Melbury Terrace
London NW1
01-262 3232

(Geography, travel, first aid, transport history)

Centre for World Development Education (CWDE)
128 Buckingham Palace Road
London SW1W 9SH
01-730 8332

(Third world studies)

Colour Centre Slides Ltd
Hilltop
Hedgerley Hill
Hedgerley, Slough
01-369 3663

(Ancient and medieval history, archaeology)

F C Curtis Ltd
6 Miletas Place
Lytham St Annes
Lancs FY8 1BQ
0253 735381

(Safety training aids)

General Dental Council
37 Wimpole Street
London W1M 8DQ
01-486 2171

(Dental health)

Geoslides
4 Christian Fields
London SW16 3JZ
01-764 6292

Hulton Educational Publications
Raans Road
Amersham
Bucks HP6 6JJ
02403 4196

Ladybird Books Ltd (distributors only)
PO Box 12
Beeches Road
Loughborough
Leics LE11 2NQ
0509 68021

Marian Ray
36 Villiers Avenue
Surbiton, Surrey
01-390 1800

(The sciences)

Mathematical Pie Ltd
West View, Fiveways
Hatton, Warcs
0926 87504

(Maths)

Multimedia Publishing Ltd
PO Box 76
Chislehurst
Kent BR7 5HL
01-467 7416

National Health Service
Learning Resources Unit
Sheffield City Polytechnic
55 Broomgrove Road
Sheffield S10 2NA
0742 661862

(Slide/tape programmes on technical aspects of nursing and programme texts)

Nicholas Hunter Ltd
PO Box 22
Oxford OX2 2JP
0865 52678

Oxford University Press
Education Department
Walton Street
Oxford OX2 6DP

(Primary English)

Philip Harris Biological
Oldmixon
Weston-super-Mare
Avon VS24 9BJ
0934 27534

Royal Society for the Prevention of Cruelty to Animals (RSPCA)
Education Dept
The Causeway
Horsham
Sussex RH12 1HG
0403 64181

(Animal welfare, zoology)

Space Frontiers Ltd
30 Fifth Avenue
Havant, Hants PO9 2PL
0705 475313

(Astronomy, space exploration)

Supervisory Management Training Ltd
21 Green Lane
London SE20 7JA
01-778 1661

Walton Sound and Film Services
Walton House
87 Richford Street
London W6
01-743 9421

Woodmansterne Ltd
Holywell Industrial Estate
Watford, Herts
92 28236

(Publishers and distributors of colour slides/audio visual material)

PRODUCERS/DISTRIBUTORS OF OVERHEAD PROJECTOR TRANSPARENCIES

Aerofilms Ltd
Gate Studios
Station Road
Boreham Wood, Herts
01-207 0666

(Geographical)

Anivis Visual Aids
The Model-making Business
Mill Street Studios
Bridgenorth,
Shropshire WB15 5AG
07462 61537

(Engineering, trigonometry, human biology, junior maths)

E J Arnold and Son Ltd
Butterley Street
Leeds LS10 1AS
0532 442944

(Map outlines, ruled music staves, grids and rulings)

Audio Visual Productions
Hocker Hill House
Chepstow
Gwent NP6 5ER
02912 5439

(History, historical maps, social studies, EFL, French, geography, the sciences)

Careers and Occupational Information Centre (COIC)
Manpower Services Commission
Moorfoot
Sheffield S1 4QP

(Careers)

W and R Chambers Ltd
11 Thistle Street
Edinburgh EH2 1DG
031-225 4463

(Modern maths course)

Educational Productions
Bradford Road
East Ardsley
Wakefield, Yorks
0924 823971

(Geography, maths, physics, chemistry)

Edu-Tex
85 St Germain's Lane
Marske-by-Sea
Redcar
Cleveland TS11 7EL
0642 477373

(Technology, engineering)

Fergus Davidson Associates (Distributors only)
1 Bensham Lane
Croydon
Surrey CR0 2RU
01-689 6824

(History, social studies, geography, biology)

Fordigraph Division
Ofrex Group Ltd
Ofrex House
Stephen Street
London W1A 1EA
01-636 3686

(English, hygiene, health, biology, history, geography, maths, sciences, creative writing, space travel, art)

Griffin and George Ltd
Gerrard Biological Centre
Worthing Road
East Preston
West Sussex BN16 1AS
09062 72071

(Sciences)

Holmes McDougall Ltd
Allander House
137-141 Leith Walk
Edinburgh EH6 8NS
031-554 9444

(Geography, sex education, human biology)

International Computers Ltd
60 Portman Road
Reading RG3 1NR
0734 595711

(Computing courses)

ITL Vufoils Ltd
10-18 Clifton Street
London EC2A 4BT
01-247 7305/8

McGraw-Hill Book Co (UK) Ltd
Shoppenhangers Road
Maidenhead
Berks SL6 2QL
0628 23431

(Business studies, office skills, accountancy, electronics)

National Computing Centre Ltd
Oxford Road
Manchester M1 7ED
061-228 6333

(Computer courses)

Pennant Audio Visual Systems
King Alfred Way
Cheltenham
Glos GL52 6QP
0242 35334

('Opasym' animated OHPs: computing, engineering, physics, electronics, motor mechanics, human anatomy)

Philip Harris Biological
Oldmixon
Weston-super-Mare
Avon VS24 9BJ
0934 413606

(Human biology, zoology, biochemistry)

Pitman Publishing
39 Parker Street
London WC2
01-242 1655

(Computer studies, engineering, physics, human anatomy, office skills)

Prentice-Hall International
66 Wood Lane End
Hemel Hempstead
Herts HP2 4RG
0442 58531

(Maths, first aid, health and hygiene, human anatomy, nursing)

Road Transport Industry Training Board
Training Resources Information Unit
Capitol House
Empire Way
Wembley HA9 0NG
01-902 8880 ext 28

Seeco
43 Store Street
London WC1E 1NR
01-637 9837

(Computing courses)

South West Regional Management Centre
Bristol Polytechnic
Coldharbour Lane
Frenchay, Bristol
0272 656261 ext 278

(Management and communications)

Transaid Visual Aids Service
Francis Gregory and Son Ltd
Spur Road, Feltham
Middlesex TW14 0SX
01-890 6455/7

Transart Visual Products Ltd
East Chadley Lane
Godmanchester PE18 8AU
0480 55251

(Education, history, maths, geography, the sciences, technology)

The Visual Aids Centre
For address see page 104

A Wheaton and Company
(Pergamon Press)
Hennock Road
Exeter EX2 8RP
0392 74121

(History, geography, sciences)

PRODUCERS/DISTRIBUTORS OF 16mm FILMS FOR EDUCATION AND TRAINING

Arts Council of Great Britain
105 Piccadilly
London W1V 0AL
01-629 9495

(Arts)

BBC Enterprises
Villiers House
Ealing Broadway
London W5
01-743 8000

British Gas Education Service
Education Liaison Officer
Room 707A
326 High Holborn
London WC1V 7PT
01-242 0789

BP Co Ltd
Britannic House
Moor Lane
London EC2
01-920 7133

British Transport Films
Melbury House
Melbury Terrace
London NW1 6PL
01-262 3232

British Universities Films Council
81 Dean Street
London W1V 6AA
01-734 3687

(Higher education)

Camera Talks Ltd
21 North Row
London W1R 2EN
01-493 2761

Central Film Library
Chalfont Grove
Gerrard's Cross
Bucks SL9 8TN
0240 74111

Concord Films Council (distributors only)
201 Felixstowe Road
Ipswich, Suffolk
0473 76012

(Social, arts)

CTVC Film Library
Foundation House
Walton Road
Bushey, Watford
Herts WD2 2JF
92 35446

(Christian, social and general interest)

Distributive Industry Training Board Film Library
Maclaren House
Talbot Road
Stretford
Manchester M32 0FP
061-872 2494

(Many industrial training boards, like DITB, have a film library)

Edward Patterson Associates Ltd
68 Copers Cope Road
Beckenham
Kent BR3 1RT
01-658 1515

(General education, specialise in humanities, health and safety)

Eothen Films Ltd
EMI Film Studios
Shenley Road
Borehamwood
Herts WD6 1JG
01-953 1600

Fergus Davidson Associates (distributors only)
1 Bensham Lane
Croydon
Surrey CR0 2RU
01-689 6824

Film Distributors Associated
Building No 9
East Lane
Wembley
Middlesex HA9 7QB
01-908 2366

(16mm films from 20th Century Fox and United Artists)

General Dental Council
37 Wimpole Street
London W1M 8DQ
01-486 2171

(Dental Health)

Guild Sound and Vision
Woodston House
Oundle Road
Peterborough PE2 9PZ
0733 63122

(All school subjects, management training, Open University films)

Hugh Baddeley Productions
64 Moffats Lane
Brookmans Park
Hatfield
Herts AL9 7RU
77 54046

(History, geography, science)

McGraw-Hill Book Co (UK) Ltd
Shoppenhangers Road
Maidenhead
Berks SL6 2QL
0628 23432

(Management training)

Mary Glasgow Publications
Brookhampton Lane
Kineton
Warcs CV35 0JB
0926 640606

(EFL, foreign languages)

Millbank Films Ltd
Thames House North
Millbank
London SW1P 4QG
01-834 4444

(Industrial relations, safety, management)

Educational Foundation for Visual Aids
National Audio-Visual Aids Centre and Library
Paxton Place
Gipsy Road
London SE27 9SR
01-670 4247/8/9

(Major distributor of all school subjects, management, training, teacher education)

National Coal Board
Film Branch
Hobart House
Grosvenor Place
London SW1
01-235 2020 ext 34855

(Training, general interest, reviews)

Open University Educational Enterprises Ltd
12 Cofferidge Close
Stony Stratford
Milton Keynes MK11 1BY
0908 566744

Post Office Telecommunications
Film Library
25 The Burroughs
Hendon
London NW4 4AT
01-202 5342

Rank Film Library
Rank Audio-Visual
PO Box 20
Great West Road
Brentford
Middlesex TW8 9HR
01-568 9222

Road Transport Industry Training Board
Training Resources Information Unit
Capitol House
Empire Way
Wembley HA9 0NG
01-902 8880 ext 284

Scottish Central Film Library
Dowanhill
74 Victoria Crescent Road
Glasgow G12 9JN
041-334 9314

(Careers series on engineering and building)

Shell Film Library
25 The Burroughs
Hendon
London NW4 4AT
01-202 7803

Stewart Film Distributors Ltd
107-115 Long Acre
London WC2E 9NU
01-240 5148

(Welding, safety)

Training Films International Ltd
St Mary's Street
Whitchurch
Shropshire SY13 1QY
0984 2597

(Industrial training)

Video Arts
2nd Floor
Dumbarton House
68 Oxford Street
London W1N 9LA
01-637 7288

(Training for industry and education)

Viewtech Audio Visual Media (Distributors only)
122 Goldcrest Road
Chipping Sodbury BS17 6XF
0272 773422

Weston Woods Studios Ltd
14 Friday Street
Henley on Thames
Oxon
04912 77033

(Children's stories)

C H Wood
500 Leeds Road
Bradford
Leeds BD3 9RV
0274 32362

(Industrial training)

PRODUCERS/DISTRIBUTORS OF VIDEOCASSETTES

Advanced Systems
364-66 Kensington High Street
London W14 8QZ
01-602 3011

(Training for computers)

BBC English by Radio and Television
PO Box 76
Bush House
Strand, London WC1
01-240 3456

(EFL)

BBC Enterprises
Villiers House
Ealing Broadway
London W5
01-743 8000

(BBC programmes)

Bell & Howell AV Ltd (Distributors only)
Alperton House
Bridgewater Road
Wembley
Middlesex HA0 1EG
01-903 5411

BP Co Ltd
Britannic House
Moor Lane
London EC2
01-920 7133

British Transport Films
Melbury House
Melbury Terrace
London NW1
01-262 3232

British Universities Films Council
81 Dean Street
London W1V 6AA
01-734 3687

(Higher education)

Camera Talks Ltd
31 North Row
London W1R 2EN
01-493 2761

(Health, hygiene, medical)

The Leonard Cheshire Foundation
Leonard Cheshire House
26-29 Maunsel Street
London SW1P 2QN
01-828 1822

Chiltern Consortium
Wall Hall
Aldenham
Watford, Herts WD2 8AT
09276 5880

Concord Films Council (distributors only)
201 Felixstowe Road
Ipswich, Suffolk IP3 9BJ
0473 76012

(Social/industrial problems)

CTVC Film Library
Foundation House
Walton Road
Bushey
Watford
Herts WD2 2JF
92 35446

Daedal Training (distributors only)
20 Green Lane
London SE20
01-778 2616

(Technical and commercial training)

Distributive Industry Training Board Film Library
Maclaren House
Talbot Road
Stretford
Manchester M32 0FP
061-872 2494

Drake Educational Associates (distributors only)
212 Whitchurch Road
Cardiff CF4 3XF
0222 24502

(In-service training for teachers)

Dundee College of Education
Gardyne Road
Broughty Ferry
Dundee DD5 1NY
0382 453433

(Teacher education)

Edward Patterson Associates
68 Copers Cope Road
Beckenham
Kent BR3 1RJ
01-658 1515

(General education, specialisation in humanities, health and safety)

Eothen Films Ltd
EMI Film Studios
Shenley Road
Borehamwood
Herts WD6 1JG
01-953 1600

Fergus Davidson Associates (Distributors only)
376 London Road
West Croydon
Surrey
01-689 6824

500 Video (Bradford) Ltd
500 Leeds Road
Bradford BD3 9RU
0274 32362

(General, industrial and commercial)

Graves Medical Audio Visual Library
220 New London Road
Chelmsford
Essex CM2 9BJ
0245 83351

(Medical and para-medical)

Guild Sound and Vision
Woodston House
Oundle Road
Peterborough PE2 9PZ
0733 63122

(Entertainment for ex-patriates)

Hugh Baddeley Productions
64 Moffats Lane
Brookmans Park
Hatfield
Herts AL9 7RU
77 54046

(Educational)

Institute of Chartered Accountants
Chartered Accountant Hall
Moorgate Place
London EC2P 2BJ
01-628 7060

Intervision Video Ltd
Unit 1
McKay Trading Estate
Kensal Road
London W10
01-960 8211

(General, sport)

IPC Video
Surrey House
Throwley Way
Sutton
Surrey SM1 4QQ
01-643 8040

(Distributors for Transworld International Films, sporting events, features, children's films, music films)

IVS Enterprises Ltd
Redan House
1 Redan Place
London W2 4SA
01-727 1556/3363

(Islam educational and Arabic feature films)

Magnetic Video UK
Perivale Industrial Park
Greenford
Middlesex UB6 7RU
01-997 2552

Mary Glasgow Publications
Brookhampton Lane
Kineton
Warcs CV35 0JB
0926 640606

(Languages, EFL)

National Audio-Visual Aids Library
Paxton Place
Gipsy Road
London SE27 9SR
01-670 4247/8/9

(School subjects, management, training, teacher education)

National Computing Centre
Oxford Road
Manchester M1 7ED
061-228 6333

Open University Educational Enterprises Ltd
12 Cofferidge Close
Stony Stratford
Milton Keynes MK11 1BY
0908 566744

Oxford Educational Resources
197 Botley Road
Oxford OX2 0HE
0865 49988/41474

(Biology, medical and health education)

Ralph Tuck Promotions Ltd
14 Stradbroke Road
Southwold
Suffolk IP18 6LQ
0502 723571

(Retail selling, personnel selection, training programmes)

Rank Audio Visual
PO Box 20
Great West Road
Brentford
Middlesex TW8 9HR
01-568 9222

(Industrial and educational)

Road Transport Industrial Training Board
Training Resources Information Unit
Capitol House
Empire Way, Wembley
Middlesex HA9 0NG
01-902 8880 ext 284

Scottish Central Film Library
Dowanhill
74 Victoria Crescent Road
Glasgow G12 9JN
041-334 9314

Shell Film Library
25 The Burroughs
Hendon
London NW4 4AT
01-202 7803

Sony UK Ltd
Pyrene House
Sunbury Cross
Sunbury-on-Thames
Middlesex PW16 7AT
09327 89581

(Video techniques)

Training Films International Ltd (distributors only)
St Mary's Street
Whitchurch
Shropshire SY13 1QY
0948 2597

also
159 Great Portland Street
London W1
01-323 1144

(Industrial relations/safety/sales)

Video Arts
Second Floor
Dumbarton House
68 Oxford Street
London W1N 9LA
01-637 7288

(Training for industry and education)

Video Inclusive
49-51 Norwood Avenue
Herne Hill
London SE24 9AA
01-674 7799

(General, documentary, music)

Videoview (London) Ltd
68-70 Wardour Street
London W1V 3HP
01-437 1333

Viewtech Audio Visual Media (Distributors only)
122 Goldhurst Road
Chipping Sodbury BS17 6XF
0272 773422

Visnews
Distribution Service Dept
Cumberland Avenue
London NW10 7EH
01-965 7733

PRODUCERS/DISTRIBUTORS OF AUDIO CASSETTES

(NB Most will supply open reel tapes, some may have records in stock)

E J Arnold and Son Ltd
Butterley Street
Leeds LS10 1AX
0532 442944

(Language, music, social studies)

Audio Arts
6 Briarwood Road
London SW4 9PX
01-720 9129

(Audio magazine of contemporary arts)

Audio Learning Ltd
Sarda House
183-189 Queensway
London W2
01-727 2748

(The sciences, social sciences, history, literature, economics, geography)

Audio Visual Education
34 Third Avenue
Hove, Sussex
0273 778161

(Computer courses)

Audio Visual Library Services
10-12 Powdrake Road
Grangemouth
Stirlingshire FK3 9UT
03244 71521

(Spoken word cassettes – English literature)

Audio Visual Productions
Hocker Hill House
Chepstow
Gwent NP6 5ER
02912 5439

(English literature)

Bible News
Mary Glasgow Publications
EAV Dept
Brookhampton Lane
Kineton
Warcs CV35 0JB
0926 640606

BBC English by Radio and Television
PO Box 76
Bush House
Strand, London WC2 4PH
01-240 3456

(EFL courses for secondary/adult)

Calibre
Wendover
Aylesbury
Bucks HP22 6PY
0296 623119

(Cassette library for the blind)

Cambridge University Press
PO Box 110
Cambridge CB2 3RL
0223 64122

(EFL, foreign languages)

Camera Talks
31 North Row
London W1R 2EN
01-493 2761

(Health, paramedical)

Chiltern Consortium
Wall Hall
Aldenham
Watford
Herts WD2 8AT
09276 5880

CTVC Film Library
Foundation House
Walton Road
Bushey
Watford WD2 2JF
92 35446

(Religious, social and general interest)

Drake Educational Associates Ltd
212 Whitchurch Road
Cardiff CF4 3XF
0222 24502

(Literature, stories for children, adult reading)

Educational Productions
Bradford Road
East Ardsley
Wakefield, Yorks
0924 823971

English by Listening
204 Raleigh House
Dolphin Square
London SW1
01-834 5323

(EFL course)

Euro-Lang Tapes
88 Wychwood Avenue
Knowle, Solihull
West Midlands
05645 4452

(Language courses for business personnel)

Fergus Davidson Associates (distributors only)
1 Bensham Lane
Croydon
Surrey CR0 2RU
01-689 6824

(Music, literature, stories for children, business management, leisure, history, arts, religion, health and sex education, Scottish performing arts)

Holmes McDougall Ltd (distributors only)
Allander House
137-141 Leith Walk
Edinburgh EH6 8NS
031-554 9444

(Children's remedial readers)

Hugh Baddeley Productions
64 Moffats Lane
Brookmans Park
Hatfield
Herts AL9 7RU
77 54046

(History, geography, science)

Institute of Chartered Accountants
Chartered Accountant Hall
Moorgate Place
London EC2P 2BJ
01-628 7060

(In-service accountancy courses)

International Computers Ltd (distributors only)
60 Portman Road
Reading, Berks
0734 595711

(In-service computer courses)

International Correspondence Schools
341 Argyle Street
Glasgow G2 8LW
Scotland
041-221 2926

(Secretarial skills, languages)

Ivan Berg Associates
35a Broadhurst Gardens
London NW6 3QT
01-624 7785

(*The Times* cassettes, *The Times* Shakespeare, literature, history, general studies, leisure)

Kiddy Kassettes
Norfolk Street Studios
Norfolk Street
Worthing
Sussex BN11 4BB
0903 205053

(Stories and music for children)

Ladybird Books Ltd
PO Box 12
Beeches Road
Loughborough
Leics LE11 2NQ
0509 68021

('Key Words' reading scheme)

Lingaphone Institute
Lingaphone House
Beavor Lane
London W6 9AR
01-741 1655

(Languages courses)

Longman Group Ltd
Longman House
Burnt Mill
Harlow
Essex CM20 2JE
0279 26721

(English, EFL, foreign languages)

Mary Glasgow Publications Ltd
140 Kensington Church Street
London W8
01-229 9531

(EFL, foreign languages, social studies, religion)

Norwich Tapes Ltd
Markfield House
Caldbec Hill
Battle, Sussex TN33 0JS
042 46 3267

(English)

Open University Educational Enterprises Ltd
12 Cofferidge Close
Stony Stratford
Milton Keynes
Beds MK11 1BY
0908 566744

(OU course material)

Oxford University Press
Education Department
Walton Street
Oxford OX2 6DP
0865 56767

(EFL, foreign languages)

Pinnacle Storyteller Cassettes
Pinnacle Electronics
St Mary Cray
Orpington
Kent BR5 3QJ
66 27099

(Stories for children)

Reach-a-Teacha Ltd
2 Hastings Court
Collingham
Wetherby
W Yorks LS22 5AW
0937 72698

(Secretarial skills, management, office safety, sales techniques)

Record and Tape Sales (distributors only)
Newmarket Road
Bury St Edmunds
Suffolk IP33 3YB
0284 68011

(Literature, stories, general interest)

Scripture Union House
130 City Road
London EC1V 2NJ
01-250 1966

(Religion, social studies)

Seminar Cassettes Ltd
218 Sussex Gardens
London W2 3UD
01-262 7357

(EFL, politics, psychology, general studies)

Stillitron
72 New Bond Street
London W1Y 0QY
01-493 1177

(Language courses)

Studio Two Educational
6 High Street
Barkway
Royston
Herts SG8 8EE
0763 84759

(Prehistoric/ancient history, industrial training)

Students Recordings Ltd
88 Queen Street
Newton Abbot
Devon
0626 3484

(English, English literature, stories for children, speech training, history, commerce, secretarial skills, economics)

Studytapes Ltd
Educational Productions
Bradford Road
East Ardsley
Wakefield
Yorks WF3 2JR
0924 823971

(English, history, sociology, foreign languages, maths, the sciences, O level +)

Sussex Tapes
Educational Productions
Bradford Road
East Ardsley
Wakefield
Yorks WF3 1BR
0924 823971

(Discussions by authorities for A level students on English literature, foreign literature, history, geography, politics, economics, science)

Talktapes
13 Croftdown Road
London NW5 1EL
01-485 9981

(Wide range of educational and spoken word cassettes, suitable for individual and group use by adults and children)

Tutor Tape Co Ltd
2 Replingham Road
London SW18 5LS
01-870 4128

(EFL, foreign languages)

University of Exeter Language Centre
Queen's Building
The Queen's Drive
Exeter EX4 4QH
0392 77911

(Foreign languages and literature, EFL)

Vistech
74 Brodrick Road
London SW17 7DY
01-672 1751

(Secretarial studies and commerce)

Waterlow (London) Ltd (distributors only)
Holywell House
Worship Street
London EC2A 2EN
01-247 5400

(Management, business studies, law)

Weston Woods Studios Ltd
14 Friday Street
Henley-on-Thames
Oxon
04912 77033

(Children's stories)

World of Learning
359 Upper Richmond Road West
East Sheen
London SW14
01-878 4314

(PILL language courses, Tee Line shortland)

PRODUCERS/DISTRIBUTORS OF 8mm FILM LOOPS

British Transport Films
Melbury House
Melbury Terrace
London NW1
01-262 3232

Camera Talks
31 North Row
London W1R 2EN
01-493 2761

Eothen Films (Int) Ltd
EMI Film Studios
Shenley Road
Borehamwood
Herts WD6 1JG
01-953 1600

Fergus Davidson Associates (Distributors only)
1 Bensham Lane
Croydon
Surrey CR0 2RU
01-689 6824

(Distributors of *Encyclopaedia Britannica* audio visual material)

John Murray Ltd
Educational Dept
50 Albemarle Street
London W1X 4BD
01-499 1792

Macmillan Educational Ltd
Houndmills
Basingstoke, Hants
0256 29242

National Audio-Visual Aids Library
Paxton Place
Gipsy Road
London SE27 9SR
01-670 4247/8/9

Rank Audio Visual
PO Box 20
Great West Road
Brentford
Middlesex TW8 9HR
01-568 9222

Viewtech Audio Visual Media (Distributors only)
122 Goldcrest Road
Chipping Sodbury BS17 6XF
0272 773422

PRODUCERS/DISTRIBUTORS OF WALLCHARTS

Adam Rouilly Ltd
Crown Quay Lane
Sittingbourne
Kent
0795 71378

(Human biology, anatomy, health)

Bank Education Service
10 Lombard Street
London EC3
01-626 9386

(Banking)

Batiste Publications Ltd
Pembroke House
Campsbourne Road
Hornsey, London N8
01-340 3291

(Human biology, botany, zoology)

Bridport-Gundry Marine Ltd
Bridport
Dorset
0308 56666

(Zoology, fish types)

British Gas Education Service
Education Liaison Officer
Room 707A
326 High Holborn
London WC1V 7PT
01-242 0789

(The gas industry)

British Museum (Natural History)
South Kensington
London SW7
01-589 6323

(Natural history)

BP Educational Service
Britannic House
Moor Lane
London EC2Y 9BU
01-920 6100

(Environmental studies, technology, engineering, the petrochemical industry, North sea, plastics, energy resources)

Butter Information Council
Pantiles House
2 Neville Street
Tunbridge Wells TN2 5TT
0892 42888

(Home economics)

Careers and Occupational Information Centre (COIC)
Manpower Services Commission
Moorfoot
Sheffield S1 4QP

(Careers guidance)

Christian Aid
PO Box 1
London SW9 8BH
01-733 5500

(Third world studies, Christian Aid activities)

Clearway Publishing Co Ltd
19 Nechells House
Dartmouth Street
Birmingham B7 4AA
021-359 2495

(Law Society charts)

Daily Telegraph
Dept CPP
135 Fleet Street
London EC4
01-353 4242

('Career plan' posters)

Educational Productions Ltd
Bradford Road
East Ardsley
Wakefield, Yorks
0924 823971

(All subjects, including *Sunday Times* wallcharts)

Electricity Council
30 Millbank
London SW1 4RD
01-834 2333

('Understanding electricity')

English Sewing Ltd
Education Dept
56 Oxford Street
Manchester M60 1HJ
061-228 1144

('Learning about sewing' charts)

European Schoolbooks Ltd
Crost Street
Cheltenham
Glos GL53 0HX
0242 45252

(Languages, European studies)

Fire Protection Association
Information and Publications Centre
Aldermary House
Queen Street
London EC4
01-248 5222

(Fire protection for industry)

Francis Chichester Ltd
9 St James' Place
London SW1Y 1PE
01-493 0931

(History)

Frederick Warne and Co Ltd
40 Bedford Square
London WC1B 3HE
01-580 9622

(Natural history)

General Dental Council
37 Wimpole Street
London W1M 8DQ
01-486 2171

(Dental health)

Griffin & George Ltd
Gerrard Biological Centre
Worthing Road
East Preston
W Sussex BN16 1AS
09062 72071

(Biology)

Globe Education
Houndmills
Basingstoke
Hants RG21 2XS
0256 29242

(Environmental studies, history, natural history, wall maps)

Health Education Council
78 New Oxford Street
London WC1A 1AH
01-637 1881

(Health education)

Help the Aged
Education Dept
218 Upper Street
London N1
01-359 0316

(Social welfare, 'Help the Aged' activities)

Hestair Hope Ltd
St Philip's Drive
Royton, Oldham, Lancs
061-633 3935

('Philips' wall maps)

Hulton Educational Publications
Raans Road
Amersham
Bucks HP6 6JJ
02403 4196

(Outline blackboard charts, maps, grids and rulings)

International Computers Ltd
60 Portman Road
Reading, Berks
0734 595711

(Computing in schools)

James Galt and Co Ltd
Brookfield Road
Cheadle
Cheshire SK8 2PN
061-428 8371

(Pictorial maps, friezes, junior topics)

D Jobson
East Wing
The Old Rectory
Wigginton
Nr York YO3 8PR
0904 761155

(Wall charts on York)

Keep Britain Tidy Group
Bostel House
37 West Street
Brighton BN1 2RE
0273 23585

(Environmental studies, 'Keep Britain Tidy' campaign posters)

London Transport Shop
280 Old Marylebone Road
London NW1
01-262 3444

(Maps, London Transport history)

Macmillan Education
Houndmills
Basingstoke, Hants
0256 29242

(History, geography)

Mathematical Pie Ltd
West View
Fiveways, Hatton, Warcs
0926 87504

(Mathematics)

National Dairy Council
Nationa Dairy Centre
John Princes Street
London W1N 0AP
01-499 7822

(Home economics, health, farming)

OXFAM
274 Banbury Road
Oxford OX2 7DZ
0865 56777

(Third world studies, activities of OXFAM)

Philip Harris Biological
Oldmixon
Weston-super-Mare
Avon VS24 9BJ
0934 27534

Pictorial Charts Educational Trust
27 Kirchen Road
London W13 0YD
01-567 9206

(All subjects)

Post Office Telecommunications
2-12 Gresham Street
London EC2 7AG
01-432 1234

(Telecommunications)

Royal Society for the Prevention of Accidents
Cannon House
The Priory
Queensway
Birmingham B4 6BS
021-233 2461

(Safety)

Royal Society for the Prevention of Cruelty to Animals (RSPCA)
Education Department
The Causeway
Horsham
Sussex RH12 1HG
0403 64181

(Animal care)

Space Frontiers Ltd
30 Fifth Avenue
Havant
Hants PO9 2PL
0705 475313

(Astronomy, space exploration)

Stanley Tools Ltd
Woodside
Sheffield S3 9PD
0742 78678

(Woodwork, use of tools)

Studio Two
6 High Street
Barkway, Royston
Herts SG8 8EE
0763 84 759

(Zoology, environmental and nature studies)

The Visual Aids Centre
78 High Holborn
London WC1
01-242 6631
Telex 268 312 Wescom G

PRODUCERS/DISTRIBUTORS OF SCIENTIFIC MODELS

Adam Rouilly (London) Ltd
Crown Quay Lane
Sittingbourne
Kent
0795 71378

(Anatomical)

Brookwick, Ward and Co Ltd
8 Shepherds Bush Road
London W6 7PQ
01-743 0875

(Anatomical)

Cochranes of Oxford Ltd
Fairspear House
Leafield
Oxford OX8 5NT
099 387 641

(Molecular)

Educational and Scientific Plastics Ltd
Holmethorpe Avenue
Redhill
Surrey RH1 2PF
91 62787

(Anatomical)

Francis Gregory and Son Ltd
Spur Road
Feltham
Middlesex
01-890 6455/7

(Model kits production)

General Dental Council
37 Wimpole Street
London W1M 8DQ
01-486 2171

(Dental health)

Griffin and George Ltd
Gerrard Biological Centre
Worthing Road
East Preston
West Sussex BN16 1AS
09062 72071

(Anatomical/botany/zoology)

Philip Harris Biological
Oldmixon
Weston-super-Mare
Somerset
0934 27534

(Anatomical and molecular)

Stuart Turner Limited
Henley-on-Thames
Oxon RG9 2AD
04912 2655

(Model steam engines)

MULTI-MEDIA KITS

Multimedia Publishing Ltd
Sales Office
PO Box 76
Chislehurst
Kent BR7 5HL
01-467 7416

(Management and staff training programmes and sales promotion)

RAW SOFTWARE

MANUFACTURERS/SUPPLIERS OF PHOTOGRAPHIC MATERIAL

Agfa-Gevaert Ltd
27 Great West Road
Brentford
Middlesex TW8 9AX
01-560 2131

Fujimix
Hanimex House
Faraday Road
Dorcan
Swindon SN3 5HW
0793 26211

Goldfinger Ltd
329 The Broadway
Muswell Hill
London N10
01-883 5502
(Mainly black and white)

Ilford Ltd
Ilford House
Christopher Martin Road
Basildon
Essex ES14 3ET
0208 27744
(No colour negatives)

J R Distributing Co Ltd
Waterside
Chesham, Bucks
0494 786331

Kentmere Ltd
Staveley
Near Kendal
Cumbria LA8 9PB
0539 821365
(Photographic paper only)

Kento Photographic Products
High Street
Kempsford
Glos GL7 4EQ
028581 426

Kodak Ltd
PO Box 33
Swallowdale Lane
Hemel Hempstead
Herts HP2 7EU
0442 61241

Konishiroku
150 Huntdon Road
Feltham
Middlesex TW13 6BH
01-751 6121

May and Baker
Rainham Road South
Dagenham
Essex PM10 7XS
01-592 3060
(Chemicals only)

Nicholas Hunter Filmstrips
PO Box 22
Oxford OX1 2JP
0865 52678

Ozalid (UK) Ltd
Langston Road
Loughton
Essex IG10 3TH
01-508 5544

Paterson Products Ltd
2-6 Boswell Court
London WC1 3PF
01-405 2826
(Darkroom materials only)

Philips Electronics
City House
420-430 London Road
Croydon
Surrey CR9 3QR
01-689 2166

Photo Technology Ltd
Cranbourne Industrial Estate
Potters Bar
Herts EN6 3JN
77 50295
(Photographic chemicals only)

Polaroid (UK) Ltd
Ashley Road
St Albans
Herts AL1 5PR
56 59191

Rank Audio Visual
PO Box 70
Great West Road
Brentford
Middlesex TW8 9HR
01-568 9222

3M (UK) Ltd
3M House
PO Box 1
Bracknell
Berks R12 1JU
0344 26726

The Visual Aids Centre
78 High Holborn
London WC1
01-242 6631

Wiggins Teape Ltd
Glory Mill
Woburn Green
High Wycombe
Berks
0494 24751
(Paper only)

MANUFACTURERS/SUPPLIERS OF OHP MATERIALS

E J Arnold and Son Ltd
Butterley Street
Leeds LS10 1AX
0532 442944

Associated Visual Products
14-20 Prospect Place
Welwyn
Herts AL6 9EN

Carbonium Ltd
Newcombe Way
Orton
Southgate
0733 234737

Chartpak Ltd
Station Road
Didcot
Berks
023581 2607

Damask
10 Liverpool Road
Blackpool
Lancs
0253 28339

Dymo Ltd
Spur Road
Feltham
Middlesex TW14 0SL
01-890 1388

Educational Foundation for Visual Aids (EFVA)
National Audio Visual Aids Centre and Library
Paxton Place
Gipsy Road
London SE27 9SR
01-670 4247/8/9

Elite Optics
354 Caerphilly Road
Cardiff CF4 4XJ
0222 63201

General Binding Company Ltd
Clivedon House
210 Old Street
London EC1V 9BE
01-250 3303

Francis Gregory and Son Ltd
Spur Road
North Feltham Trading Estate
Feltham
Middlesex TW14 0SX
01-890 6455/7

Ofrex Ltd
Ofrex House
Stephen Street
London W1A 1EA
01-636 3686

Rexel Ltd
Gatehouse Road
Aylesbury
Bucks HP19 3DT
0296 81421

G H Smith and Partners
Berechurch Road
Colchester
Essex
0206 48221

Staedtler (UK) Ltd
Pontyclun
Mid-Glamorgan CF7 8YJ
0443 222421

Swan-Stabilo Ltd
71 Parkway
London NW1 7QJ
01-267 3512

3M (UK) Ltd
3M House
PO Box 1
Bracknell
Berks R12 1JU
0344 26726

Transart Visual Products Ltd
East Chadley Lane
Godmanchester
Huntingdon PE18 8AU
0480 51171

MANUFACTURERS/SUPPLIERS OF AUDIO TAPES

Agfa-Gevaert Ltd
Great West Road
Brentford
Middlesex TW8 9AX
01-560 2131

Ampex Great Britain Ltd
Acre Road
Reading
Berks RG2 0QR
0734 85200

BASF United Kingdom Ltd
Audio Video Tapes Division
4 Fitzroy Street
London W1P 6ER
01-388 4200

Bradbury Electronics
119A Loverock Road
Reading
Berks RG3 1NS
0734 52434

Educational Foundation for Visual Aids (EFVA)
National Audio Visual Aids
Centre and Library
Paxton Place
Gipsy Road
London SE27 9SR
01-670 4247/8/9

Fuji
Pyser Ltd
Fircroft Way
Edenbridge
Kent TN8 6HA
0732 864111

Grundig International Ltd
42 Newlands Park
Sydenham
London SE26 5NQ
01-659 2468

Hitachi (UK) Ltd
Hitachi House
Station Road
Hayes
Middlesex UB3 4DR
01-848 8787

J R Distributing Ltd
Waterside
Chesham
Bucks
0494 786331

Leeholme Audio Services Ltd
350-4 Lea Bridge Road
Leyton
London E10 7LD
01-556 4748

Maxwell
1 Tyburn Lane
Harrow
Middlesex
01-423 0688

Memorex (UK) Ltd
96-104 Church Street
Staines
Middlesex
81-51488

Musonic
S and B Trading
Stylus House
34-38 Verulam Road
St Albans
Herts AL3 4DF
56 50611

National Panasonic Ltd
300-318 Bath Road
Slough SL3 6JB
0753 34522

Olympus Optical Co (UK) Ltd
2-8 Honduras Street
London EC1
01-253 2772

Philips Audio Ltd
City House
420-430 London Road
Croydon
Surrey CR9 3QR
01-689 2166

Professional Tapes Ltd
329 Hunslett Road
Leeds LS10 1NJ
0532 706066

Revox
F W O Bauch Ltd
49 Theobalds Road
Borehamwood
Herts WD6 4RZ
01-953 0091

Ross Electronics
49-53 Pancras Road
London NW1 2QB
01-278 6371

Scotch
3M House
PO Box 1
Bracknell
Berks
0344 58357

Sony (UK) Ltd
Pyrene House
Sunbury Cross
Sunbury-on-Thames
Middlesex PW16 7AT
09327 89581

Syncrotape
Adastra Electronics Ltd
Unit N22
Cricklewood Trading Estate
Claremont Road
London NW2 1TU
01-452 6288

Tandberg (UK) Ltd
Revie Road
Elland Road
Leeds
West Yorks LS11 8JG
0532 774844

TDK
TDK Tape Distributors (UK) Ltd
Eleventh Floor
Pembroke House
Wellesley Road
Croydon
Surrey
01-680 0023

Technics
National Panasonic Ltd
300-18 Bath Road
Slough
Berks SL3 6JB
75 34522

Teletape Ltd
33 Edgeware Road
London W2
01-723 1942

3M (UK) Ltd
3M House
PO Box 1
Bracknell
Berks R12 1JU
0344 26726

The Visual Aids Centre
78 High Holborn
London WC1
01-242 6631

Yorke, James Ltd
Oak House
High Street
Northleach
Gloucestershire GL54 3DH
04516 509

MANUFACTURERS/SUPPLIERS OF VIDEO TAPES

Agfa-Gevaert Ltd
27 Great West Road
Brentford
Middlesex TW8 9AX
01-560 2131

Akai UK
Unit 12
Hazlemere Heathrow Estate
Silver Jubilee Way
Hounslow
Middlesex TW4 6NF
01-897 7171

Ampex Great Britain Ltd
Acre Road
Reading RG2 1QR
0734 864121

BASF (UK) Ltd
Video Audio Tape Division
4 Fitzroy Square
London W1P 6ER
01-388 4200

Bell & Howell Ltd
Alperton House
Bridgewater Road
Wembley
Middlesex HA0 1EG
01-903 5411

Bradbury Electronics Ltd
119A Loverock Road
Reading RG3 1NS
0734 52434

Fantasy Factory Video Ltd
42 Theobalds Road
London WC1X 8NW
01-405 6862

Ferguson Video Star
284 Southbury Road
Middlesex
01-363 5353

Fuji
Bell & Howell AV Ltd
Alperton House
Bridgewater Road
Wembley
Middlesex HA0 1EG
01-902 8812

Hitachi (UK) Ltd
Hitachi House
Station Road
Hayes
Middlesex UB3 4DR
01-848 8787

In Video (Scotland) Ltd
30A St Colnes Street
Edinburgh EH3 6AA

J R Distributing Ltd
Waterside
Chesham
Bucks
0494 786331

JVC (UK) Ltd
Eldonwall Trading Estate
Staples Corner
London NW2 7AF
01-450 2621

Memorex (UK) Ltd
96-104 Church Street
Staines
Middlesex
01-570 7716

National Panasonic Ltd
Video Department
300-318 Bath Road
Slough SL3 6JB

Philips Video Ltd
City House
420-430 London Road
Croydon
Surrey CR9 3QR
01-689 2166

Pyral (UK) Ltd
Airport House
Purley Way
Croydon CR0 OXZ
01-681 2833

Robian Video
43 King Street
Stanford-le-Hope
Essex SS17 0HJ
03756 43777

Sony Broadcast Ltd
City Wall House
Basing View
Basingstoke
Hants RG21 2LA
0256 55011

TDK Tape Distributors (UK) Ltd
11th Floor
Pembroke House
Wellesley Road
Croydon CR0 9XW
01-680 0023

Teletape Ltd
33 Edgeware Road
London W2
01-723 1942

3M (UK) Ltd
3M House
PO Box 1
Bracknell
Berks R12 1JU
0344 26726

Transart
East Chadley Lane
Godmanchester
Huntingdon
Cambridgeshire PE18 8AU
0480 51171

Video Inclusive
49-51 Norwood Avenue
Herne Hill
London SE24 9AA
01-674 7799

The Visual Aids Centre
78 High Holborn
London WC1
01-242 6631

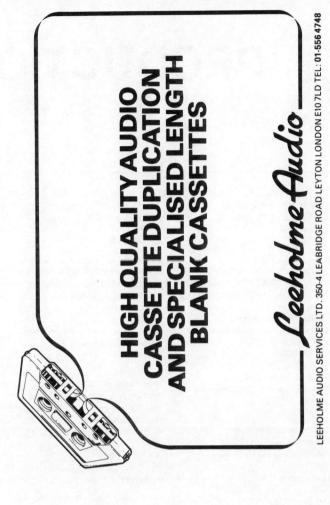

PRODUCTION SERVICES

Airtime Productions Ltd
50 Frith Street, London W1V 6PJ
01-734 9304

Specialise in origination on 2 inch broadcast standard for commercials, corporate image, etc. Also non-broadcast standard test studio on premises for hire.

Applied Audio Visual Ltd
Bridges Mews, Bridges Road, London SW19 1EL
01-540 0651

Audio + Video Ltd
Video House, 48 Charlotte Street, London W1P 1LX
01-580 7161

International standards conversion and video tape duplication facilities.

Audiogenics Ltd
34-36 Crown Street, Reading, Berkshire RG1 2SN
0734 595647

Audio visual and soundtrack production, jingles and music production. Cassette pulsing and cassette duplication.

Audio Impact
40 Quarrendon Street, London SW6 3SU
01-731 1719

Tailor-made industrial and commercial training programmes in slide/tape and filmstrip. Duplication service and slide library.

Audio Visual Material Supply Co
AVM House, 1 Alexandra Road, Farnborough, Hampshire GU14 6BU
0252 40721/3

Complete or part service in slide/tape and video production. Studio facilities (photographic and recording). Equipment hire.

Audio Visual Programmes Ltd
Studio AVP, 82 Clifton Hill, London NW8 0JT
01-624 9111

Comprehensive audio visual and conference service including programme production in foreign languages. Transfer programmes to film, filmstrip and video tape. Four-track recording studio facility and a conference room.

Audio Visual Techniques
86 Preston Road, Brighton BN1 6AE
0273 561126

Producers of multi-image audio visual shows. Full AVL programming facilities, equipment hire and recording studio.

AVE Ltd
73 Surbiton Road, Kingston-on-Thames, Surrey KT1 2HG
01-549 3464

Provide a complete range of slide/tape audio visual production services from major conference staging to a simple budget single screen programme. Rental facilities and staging capability.

AV Graphics
76 Cleveland Street, London W1
01-637 4877

Create and conceive audio visual slide/tape programmes.

AVM Video
Unit 26, Northfield Industrial Estate, Beresford Avenue, Wembley, Middlesex
01-902 9707

Facilities house with 600 sq ft studio, 2 colour cameras and a mixer. Equipment hire. Post production: transfer 35mm or 16mm film on to any type of video cassette. Duplicate tapes using U-matic as a mastering machine.

Hugh Baddeley Productions
64 Moffat Lane, Brookmans Bank, Hatfield, Hertfordshire AL9 7RU
77 54046

16mm films and slide/tape productions. Filmstrips and ¼ inch tape sound cassettes. Specialise in educational films.

Beethoven Street Presentations
25 Beethoven Street, London W10 4LG
01-960 1414

Producers for communications, conferences and presentations. All aspects of audio visual work, conference setting, design and construction. Complete presentation and conference servicing.

Benedict-Lindand Ltd
1-2 Newman Passage, Newman Street, London W1G 3HG
01-637 7809

Video production and equipment hire company.

Boulton-Hawker Films Ltd
Hadleigh near Ipswich, Suffolk IP7 5BG
0473 822235

Film and video production of all types and complexities.

Pete Brady Productions
Huffwood Trading Estate, High Street, Partridge Green, Sussex
0403 710885

Audio visual, conference and film production company

British Super 8
46 High Street, Langford, Biggleswade, Bedfordshire SG18 9RU
0462 700075

British Transport Films
Melbury House, Melbury Terrace, London NW1 6LP
01-262 3232

Audio visual presentations from most simple to multi-media, including lengthy documentaries.

Alan Brockway Ltd
26 West Street, London WC2
01-836 6018

Production of slides and filmstrips from any size original material.

Bryanston Audiovision
2 Portman Mews South, Portman Street, London W1H 9AU
01-493 1541

Multi-media kit, filmstrip, slide, printed matter production service.

Cadbury Schweppes Ltd
Bournville Lane, Birmingham B30 2LU
021-458 2000 ext 661

Multi-screen presentations for conferences etc. Cine projection in 16mm and 35mm; video programmes. Sound systems and lighting rigs available.

Cairnes Design Associates Ltd
315 New Kings Road, London SW6 4RF
01-731 2621

Software company. Video and audio visual multi-screen, writing and direction. Graphics and back-up.

Cal Video Ltd
22 High Street
Hampton Hill
Hampton
Middlesex
01-943 1272

Complete video production facilities from scripting through to duplication and distribution, including our own in-house designer. Mobile equipped with Hitachi SK19s and Sony BVU 110s. Four machine computer controlled editing, voice overs, music and FX library. Facility-hire only, if required.

CAV
36 North West Thistle Street Lane, Edinburgh EH2 1EA
031-225 8871

Produce slide/tape programmes

Chiltern Consortium
Wall Hall College
Aldenham
Watford WD2 8AT
09276 5880

Educational TV unit; programmes for in-service teacher training and higher education. Mainly video; some slide/tape sequences, audio programmes and slide sets. Produce non-broadcast programmes for colleges of higher education and local education authorities. Special emphasis on material for initial and in-service teacher training. Programmes distributed through Chiltern Resources Library.

Cinema Workshop Sales Ltd
29 Greenford Avenue, Hanwell, Ealing, London W7 1LP
01-567 4543

Cinephoto Film Productions Ltd
Cinephoto House, 17 The Crescent, Salford M5 4PF
061-736 6221/2

Film producers of industrial documentaries. Also audio visual presentations and video productions. Facilities include a dubbing theatre, editing rooms, sound stage.

Cinevideo Ltd
245 Old Marylebone Road, London NW1 5QT
01-724 1363

Full video production facilities including crewing equipment and post production facilities. Hire out video equipment.

Colour Sound Filmstrips
59 Bridge Street, Manchester M3 3BQ
061-832 2310

Comcen Communications Centre Ltd
325 City Road, London EC1V 1LJ
01-278 7831

Sound recording facilities, 35mm software production facilities.

also at:
Bledisloe, Coates, Cirencester, Gloucestershire
028577 564

Full in-house training centre, catering for all aspects of production.

Creative Film Makers Ltd
Pottery Lane House, 34a Pottery Lane, Holland Park, London W11 4LZ
01-229 5131

Sponsored documentary films, advertising films, TV commercials and audio visual presentations (slide/tape, filmstrip and video). All in-house.

Crossfade Productions
10-11 Rathbone Place, London W1P 2DN
01-636 1085

Slide/tape production team producing quality audio-visual packages. Specialise in visual effects from twin projectors on a single screen. Also transfer slide/tape to video.

CST Training Resources Ltd
Bushey Studios, Melbourne Road, Bushey, Hertfordshire WD2 3LN
01-950 1621

Specialise in production of audio visual programmes on training and safety. Design, produce and distribute programmes on video, film and slide/tape with full supporting material. Own studios and creative and technical staff.

CTVC
Hillside, Merry Hill Road, Bushey, Hertfordshire WD2 1DR
01-950 4426/7

Productions on film, video and sound cassettes. Extensive studio facilities for hire. Also training in radio and television presentation techniques. Facilities available for transfer for 16mm film to video-cassette.

Dale Hire
16 Bermondsey Trading Estate, Rotherhithe New Road, London SE16
01-231 2050

Complete conference staging facilities, including all hardware, audio visual equipment and video equipment.

Damask
10 Liverpool Road, Blackpool, Lancashire
0253 25339

OHP production services, black-and-white and colour. Artwork. White plastic re-usable frames.

R E Davis
36 St Botolph's Street, Colchester, Essex CO2 7EA
0206 73444

Produce 35mm audio visual slide programmes including multi-vision.

Peter Dearden Ltd
The Old Courthouse, The Green, Scorton, Richmond, North Yorkshire
0748 811582

Design Systems
1 Talbot Street, Cardiff CF9 1BW
0222 31231

Write and produce programmes for slide/tape and video.

Diafade Ltd
10 Cowbridge, Hertford SG14 1PQ
0992 552383

Production of slide/tape programmes.

Distributive Industry Training Board
Parkgate Estate, Knutsford, Cheshire WA16 8DX
0565 52871

Produce training programmes within the distributive industry both pre-recorded and to specification. Slide/tapes, Sony U-matic cassettes, VHS cassettes, Philips cassettes, 16mm film and OHP transparencies. Studio hire and crews. Three 2-inch quad RCA video tape recorders.

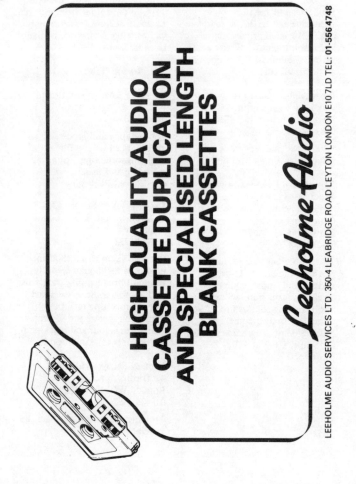

Drake Video Services
212 Whitchurch Road
Cardiff CF3 3XF
0222 24502

All video and audio visual facilities.

Edco Reed International Productions Ltd
Broadway Chambers,
14-26 Hammersmith Broadway,
London W6
01-741 1921

All facilities for full programme production including photography, script-writing, artwork, storyboard, sound-track, video, film and filmstrip production. In-house Forax rostrum camera available for 35mm, 46mm and 16mm slide production including masks, multi-exposures, split-screens and other special effects.

Edinburgh Film Productions
Nine Mile Burn, Penicuik,
Midlothian EH26 9LX
0968 72131

16mm and 35mm film and video production. Sound transfer from tape to film. Facilities include animation studio, sound stage (1333 sq ft), 100 seat preview theatre for 16mm and 35mm. Film and video cassette, cutting room.

Editorial and Production Services Ltd
240 Kentish Town Road, London
NW5 2DD
01-485 8215

Production to specification in slide/tape and OHP transparencies. Associated printed matter and integration of printed matter with programmes. Packaging of audio visual kits.

Edric Audio Visual Ltd
Oak End Way, Gerrards Cross,
Buckinghamshire SC9 8BR
02813 84646

Film, video and tape/slide production. Equipment hire.

Educational Foundation for Visual Aids (EFVA)
National Audio Visual Aids
Centre and Library
Paxton Place, London SE27 9SR
01-670 4247/8/9

700 sq ft colour video studio (CY8800 cameras to U-matic), small sound studio, equipment hire and production facilities. Short training courses in the use of audio visual equipment and production of programmes (video, slide-tape, OHP and film projectors).

Educational Productions Ltd
Bradford Road, East Ardsley,
Wakefield, West Yorkshire
WF3 2JN
0924 823971

Produce filmstrips, slide/tapes, video tapes, tape recordings, and OHP transparencies for education and industrial training. Both to specification and pre-recorded. Also wall charts, study kits and books.

Electronic Picture House
191 Wardour Street, London W1
01-439 9701

U-matic editing, cassette duplication, tele-cine film to tape transfer 8mm, 16mm and 35mm. Also equipment hire.

Enlightenment Audio Visual Productions
St Catherine's, Botes, Diss,
Norfolk IP22 1BP
037-989 434

Full production facilities including scripting, for slide/tape and film-strip production and video and film production. Also slide presentations, slide duplication and OHP transparencies. Tailor-made seminars on effective audio visual presentation.

EOS Electronics AV Ltd
EOS House, Western Square,
Barry, South Glamorgan
CF6 7TY
0446 741212

U-matic editing; audio cassette duplication; 16mm Steenbeck film editing. Facilities for hire; mobile van with 2 tele cameras, U-matic recorder, Sony, JVC.

ESL Bristol Training Services Division
Waverley Road, Yate, Bristol
BS17 5RB
0454 316774

Photography and processing to make film slides and filmstrips including mounting. Sound recording and cassette duplication. Packaging.

Express Design Service Ltd
1a Kilburn High Road, London
NW6 5SE
01-328 3232

Produce audio visual slide/tape presentations. Sound recording studio of broadcasting standards. Full photographic services and colour processing facilities. Projection theatre for 50 people. Also giant colour and black-and-white prints for exhibitions.

Fantasy Factory Video Ltd
42 Theobalds Road
London WC1X 8NW
01-405 6862

Training, documentaries, art and general video.

Fidelity Films Ltd
34-36 Oak End Way, Gerrards
Cross, Buckinghamshire SL9 8BR
49 84646/8

Make complete audio visual programmes to client specification. In-house scripting, photography and sound studio.

Film Facilities Magnetic Ltd
3 Springbridge Mews,
Springbridge Road, Ealing
Broadway, London W5
01-567 3613

Post production services: magnetic striping, sound transfer, edge numbering, waxing, scratch removal, academy leaders.

Filmstrip Services Ltd
53 Theobald Street,
Borehamwood, Hertfordshire
WD6 4RT
01-953 6077

Convert any material supplied to slide sets or filmstrips (half or full frame). Numbering in picture area; superimpose lettering on slides. Mounting in plastic, glass or card. Supply plastic wallets or boxes.

Fisher Audio Visual
Audio Visual Photographic
Equipment, 724 Borough Road,
Birkenhead, Merseyside L42 9JE
051 608 4848

Produce programmes to specifications in slide/tape, video and OHP transparencies. Hire and service equipment.

500 Video (Bradford) Ltd
500 Leeds Road
Bradford BD3 9RU
0274 32362

Mainly commercial; own video theatre slide/tape sequencies; audio productions; 16mm films.

Focus Films (Leeds) Ltd
281 Meanwood Road, Leeds
LS7 2JR
0532 625267

Film production company for TV, industry, commercials, sales, training, prestige documentaries. Broadcast video facilities. Editing and viewing. All own equipment. Hire out crews and equipment.

Frameline Productions Ltd
32 Percy Street, London
W1P 9FG
01-636 1303

Film and video production and transfer facilities.

Foundation Presentations Ltd
50 Warwick Street, Leamington
Spa, Warwickshire CV32 5JS
0926 34527

Comprehensive production capability in multi-vision, slide/tape, film, video, OHV and print. Produce conferences, training and promotional programmes, corporate communications, etc. Complete service from initial concept through to packaging and distribution.

Frameline Productions Ltd
32 Percy Street, London
W1P 9FG
01-636 1303

Rostrum camera facilities company with 16mm and 35mm rostrum cameras. Art dept producing titling, graphics and insert animation.

Fraser Peacock Associates Ltd
94 High Street, Wimbledon
Village, London SW19 5EG
01-947 7551

Audio visual programmes, especially for conferences. Cassette duplication.

John Garbett (Audio Visual) Ltd
1 Rectory Road, Wokingham,
Berkshire RG11 1DJ
0734 790415

Full in-house production capability in all media. Graphic artists, photographers and film cameramen. Also design and install audio visual systems for board rooms, conferences, etc.

Gateway Audio Visual Ltd
470-472 Green Lanes, Palmers
Green, London N13 5XF
01-882 0177

Produce documentary films for industry. Also video and slide/tape and training programmes.

Geoslides
4 Christian Fields, London
SW16 3JZ
01-764 6292

Produce to order single slide sets for lecturers and teachers, using own library material.

Brian Gibson Productions Ltd
Old Court House, Old Court
Place, Kensington, London
W8 4PD
01-937 8100

Produce sponsored documentaries, TV commercials, audio visual slide/tape and video.

Gordon AV Ltd
28/30 Market Place, Oxford Circus, London W1N 8PH
01-580 9191

Transfer of sound/slide programmes and video to super 8 format.

Gordon Hammond Audio Visual Co Ltd
97 Alma Road, Southampton, Hampshire
0703 35575/6

Slide/tape programmes, film programmes, and video programmes. Organise and operate conferences. Hire out equipment.

Francis Gregory and Son Ltd
Spur Road, Feltham, Middlesex TW14 0SX
01-890 6455/7

Print OHP transparencies according to specification.

Guardian Business Services Ltd
119 Farringdon Road, London EC1
01-278 2332

Management training consultants. Design and produce audio visual training materials and aids to client specifications. Also public courses in the use of audio visual (eg how to make slide/tape programmes).

Guild Productions (Management Training) Ltd
Woodston House, Oundle Road, Peterborough PE2 9PZ
0733 63122

Producers of all audio visual media formats for industrial and commercial training.

Colin Hamp Group Ltd
110 Clarendon Road
London W11 2HR
01-727 4269

Mainly training: industrial and general. Sound recordings; 16mm films.

Gye Handley Associates
9 Dean Street, London W1
01-439 8705

Audio visual production of slide/tape programmes ranging from single screen modules to multi-vision shows for product launches and conferences. Entirely in-house operation.

Hamper & Pursell Ltd
91-93 Gray's Inn Road, London WC1X 8TX
01-404 0033

Audio visual writing, photography and production. Multi-screen and multi-media production.

H and H Productions
Springbank, 65 East Kilbride Road, Busby, Glasgow G76 8HX
041 644 3103

Complete production facilities for tape/slide programmes and audio tapes of all types with special emphasis on educational and training sequences. Individualised learning material, including reinforcing and extension print materials also prepared. OHP transparencies produced.

Hedges Wright (AV) Ltd
40 Victoria Road, Swindon, Wiltshire SN1 5QZ
0793 31032

Comprehensive range of audio visual production from slides, slide/tape, films and video.

Humphries Video Services Ltd
32a Eveline Road, Mitcham, Surrey
01-640 5515

Video duplication house. Facilities include: Rank Cintel Mk3, Flying Spot Scanner, 35mm and 16mm Telecine, 2 inch Quad Ampex VTR sepmag facility, slide scanner and captioning. Complete service including labelling and packaging.

ICP Films Ltd
The Studio Centre, Ackhurst Road, Chorley, Lancashire PR7 1BG
02572 66411

Complete audio visual services in all media from initial meeting through to final product. Basic stills photography for posters and brochures, etc. Sound-track facilities and script-writing.

Illustra Films Ltd
Illustra House, 13-14 Bateman Street, London W1V 6EB
01-437 9611

Produce industrial documentaries, sales training, company films and commercials.

Independent Television
ITN House, 48 Wells Street, London W1P 4DE
01-637 2424

Commercial production in video tape. Pop promotions. Industrial and training programmes in video. Cassette duplication.

Infovision
Bradley Close, White Lion Street, London N1 9PN
01-837 0012

Industrial TV, film and slide/tape. Own producers, directors, writers, graphics. Production facilities: EFP unit based Ikegami and Sony camera and Sony BVU editing.

In Video (Scotland) Ltd
3a Coates Place
Edinburgh EH3 7AA
031-225 9524

Mainly industrial and commercial. Facilities for editing and transferring from one media to another.

Istead Audio Visual
38 The Tything, Worcestershire WR1 1JS
0905 29713/4

Programme production, video tape duplication and transfer, video tape electronic editing, 16mm and 8mm film to video copying, 35mm slides to video copying.

ITL Vufoils Ltd
109-123 Clifton Street, London EC2
01-739 5801

JB Presentations Ltd
15 Brackenbury Road, London W6 0BE
01-749 6063

Slide/tape producers from single screen to multi-screen conferences.

J D Studios Ltd
1 Cheniston Gardens Studios, London W8 6TL
01-937 5028

Audio visual productions, film-strips, film and video.

Kadek Vision Ltd
Shepperton Studio Centre, Studios Bridge Road, Shepperton, Middlesex TW17 0QD
093 28 66941

Complete production facilities for audio visual slide/tape, 1 to 56 projectors. AVL Electrosonic and Auvitec equipment. Total sound facilities. Also hire and sell audio visual equipment.

Kent Video
Ramillies House, Ramillies Street, London W1
01-734 5684

Produce films, slides and slide/tapes for training and education.

Kestrian International
Film Centre, James Street, Blackburn, Lancashire BB1 6BE
0254 60606

16mm short film producers for industry (sales, training, PR, documentaries). Foreign language versions (even Mandarin Chinese).

Lane End Productions Ltd
9 Meard Street, London
W1V 3QH
01-734 0072

Video and film editing facilities. Hire of video equipment.

Lee International AV Ltd
128 Wembley Park Drive,
Wembley, Middlesex HA9 8JE
01-902 3616

Audio and slide/tape production. Slide/tape, film and video transfer facilities. Equipment hiring.

Livesey Audio Visual
95 Princes Avenue
Hull
North Humberside HU5 3QR
0482 43453

Have full production facility, especially geared to producing video and slide programmes for the new portable Impact Video Sales Presenter. Equipment hiring. Customised carrying case.

London Video Ltd
Marvic House, Bishops Road,
London SW6 7AD
01-385 8443

Make video programmes to client specifications including scripting, etc, especially for industrial training and servicing.

Marall-Smith Films
15a The Mall, Ealing, London
W5 2PJ
01-567 9032/9065

Transfer slide/tape shows (multi-screen or single screen) to film. Make foreign language versions. Transfer to any format of film or video cassette. Make film inserts also, if desired. Camera and editing facilities available.

Mark Recordings Ltd
Ahed House, Sandbeds,
Dewsbury Road, Ossett,
Yorkshire WF5 9ND
0924 278181

Specialise in audio visual presentation. Slide/tape production, eg for training or promotion. Slide duplication. Multi-lingual narration and preparation of sound-tracks.

Martak Ltd
Brook House, 25 High Street,
Alton, Hampshire GU34 1EN
0420 88465

Martak Northern Ltd
Unit 12, Knutsford Industrial
Estate, Parkgate Lane, Knutsford
WA16 8DX
0565 54655

Martak Southern Ltd
86-94 High Street, Alton,
Hampshire GU34 1EN
0420 83058

Twin projector, single screen slide/tape production. Also video productions.

Media Facilities (Scotland) Ltd
3 Jamaica Street, Edinburgh
EH3 6HH
031-225 1910

Provide JVC CY8800 3 tube colour camera: U-matic editing, 1300 sq ft studio, broadcast facilities. TV programme productions including scripting, production and post production techniques.

Middlesex Film Productions Ltd
Shepperton Studio Centre,
Studios Road, Shepperton,
Middlesex TW17 0QD
093 28 60221

Full in-house audio visual production, all formats of filmstrip and slide/tape. Film production facilities especially for medical, industrial and educational training films.

Midland (AV) Ltd
Rear of 202 New Road,
Birmingham B45 9JA
021-453 3141

Produce films, documentaries, slide/tape programmes and video. Sound recording, etc. Everything in-house.

Millbank Films Ltd
Thames House North, Millbank,
London SW1P 4QG
01-834 4444

Produce industrial and management training films to client specification.

Molinare Video Ltd
43 Fouberts Place, London W1
01-439 7631

Studio facilities for TV, radio and audio visual. TV studio and mobile unit. Comprehensive computer video tape editing facilities. 'Squeezoom' for 144 effects. Chyrou IV character generator. Four camera mobile units plus EFP.

Morgan Associates
12 Manderley Close, Eastern
Green, Coventry
0203 467456

Slide/tape presentations, 35mm and Superslide. OHP transparencies. Video presentations in all formats.

MST Presentations
42 Newman Street, London W1
01-580 2566

Software production company for any type of programme using slide/tape or OHP transparencies. Video from pre-taped slide programmes.

Multimedia Publishing Ltd
Sales Office
PO Box 76
Chislehurst
Kent BR7 5HL
01-467 7416

Management and staff training programmes and sales promotion. Multimedia format.

Multi-Vision Audio Visual Ltd
Wellington House, Duke Street,
Manchester M3 4NF
061-834 9994

Slide/tape production company. Multi-screen programme facilities using AVL Eagle system. Anything from single projector to multi-vision screen presentations.

Myriad AV Sales Ltd
10/11 Great Newport Street,
London WC2H 7JA
01-240 1941

Audio visual productions, photography and sound production. Slide to film and video transfer. Equipment hire.

Richard Need Productions Ltd
107-115 Long Acre, London
WC2E 9NT
01-240 5148

Produce 16mm films on training, education and instructional techniques. Produce slide/tape presentations.

NSR Ltd
3 Parr Road, Stanmore,
Middlesex HA7 1PZ
01-951 0222/5

Recording studio facilities, programming services. Hire and sell equipment.

Opsis Ltd
83 Prospect Road, Southborough,
Tunbridge Wells, Kent TN4 0EE
0892 42708

Video production, audio and slide/tape and hire facilities.

Pantheon Film Productions Ltd
38 Mount Pleasant, London
WC1X 0AP
01-278 7621

Film, filmstrip, tape/slide, video, multivision, back projection, front projection, cross fade, register mount, multi media etc.

Ann and Bury Peerless
22 Kings Avenue, Minnis Bay,
Birchington, Kent CT7 9QL
0843 41428

Slide sets and studies, specialising in world religions but covering general aspects including art, architecture, history, geography.

Pennant Trainers and Simulators Ltd
King Alfred Way, Cheltenham,
Gloucestershire GL52 6QP
0242 35334

Screen printing, technical illustration, systems design.

Photomedia Audio-Visual Ltd
6 St Helen's Gardens, London
W10 6LR
01-960 0438

Colour slide production 35mm and 46mm format, slide copying and filmstrip production and special effects.

John Portsmouth Associates Ltd
The Coppings, Kingswood Road,
Bromley, Kent BR2 0HQ
01-460 8517

Production of slide/tapes (16mm and Super 8). Complete service from storyboard through to end product in either cassette form or reel-to-reel.

Post Sound Ltd
2 Springbridge Mews,
Springbridge Road, Ealing,
London W5
01-567 7036

Produce audio visual programmes for educational and industrial clients. Studio (small voice) with separate production channel. Pulsing facilities for most formats. High speed duplicating plant for cassette and open reel format and cartridges.

Prater Audio Visual Ltd
139 Greenwich High Road,
London SE10
01-858 5715

Programme producers. Every facility in-house: photographic studio, rostrum cameras, voice recording, mixing, dubbing and programming.

Provideo
98 Wardour Street, London W1
01-439 8901

Facilities for shooting on high grade U-matics, U-matic editing facilities using convergence editor, small studio suitable for interview and lectures with 2 IVC cameras, Cox 30X vision mixer with Chroma key and effects. Film to tape facilities with Quad VTR mastering if required.

Purchasepoint Group
2/14 Shortlands, London W6 80J
01-741 1121

Offer conference production, audio visual production, promotions and incentives. Exhibition design, graphic design, packing and print, photographic services. Also colour processing and studio in Manchester (Colour 061).

QAV Presentations Ltd
7 Guildford Road, Woking,
Surrey
048 62 70204

Create business presentations in all audio visual media, eg for exhibitions, corporate image presentations, training and sales aids.

Rank Film Laboratories
North Orbital Road, Denham,
Uxbridge, Middlesex UB9 5HQ
089583 2323

Slide/tape, video and filmstrip production. Telecine transfers.

Rank Xerox Copy Bureau
Middlesex House, 4 Mercer Walk,
Uxbridge, Middlesex UB8 1UD
0895 38230

Copying on to film in full colour for OHP from flat artwork or from 35mm transparencies.

Research Recordings Ltd
5-8 Clarendon Terrace, Edgware
Road, London W9 1BZ
01-289 2263

Transfer slide/tape programmes to video using a broadcast quality camera and high gain screen. Telecine Rank Cintel Mk3.

Right Angle Productions Ltd
Cinema House, 93-95 Wardour
Street, London W1
01-437 3962

Complete service in film and video production.

Robian Video
43 King Street
Stanford-le-Hope
Essex SS17 0HJ
03756 43777

Production facilities, location filming, editing, special effects, duplication, transfers and tele-line conversions. Everything you need for non-broadcast video programmes.

Roundel Productions Ltd
51 Loudoun Road, London
NW8 0DL
01-624 6080

Film, video, slides, tape and tape/slide productions.

Sanderson Vere Crane (Video Tape Editors) Ltd
88 Wardour Street, London
W1V 3LF
01-734 1600

Small colour TV studio (1 camera) for video programme production, editing suite, telecine, slide/tape facilities. Produce 16mm and 35mm films.

Sarner Audio Visual
32 Woodstock Grove, Shepherds
Bush, London W12
01-743 1288

Technical facilities house: recording studios, AV programming. Systems design and consultancy and multi-screen. Equipment sales and rental.

Saville Audio Visual Ltd
Salisbury Road, York YO2 4YW
0904 37700

Full slide/tape and multi-screen programme production, 16mm film production, non-broadcast video programme production. In-house sound studio and complete production suite.

SAV Presentations Ltd
132 Tanners Drive, Blakelands,
Milton Keynes MK14 5BR
0908 612586

Complete slide/tape and video film production service. Also flexible part service if required. Facilities: photographic studio, 1000 sq ft, drive-in recording studio with rear projection facility screen. Programming room and all audio visual equipment.

SAV Studios Ltd
34 Cricklewood Broadway,
London NW2 3CT
01-278 7893

Recording studios in which sound-track is programmed to visuals. Also music and effects.

SB Modules Ltd
159 Great Portland Street,
London W1N 5FD
01-323 1144

Produce slide/tape programmes and video programmes.

Screen Media Programming Ltd
3 Attenborough Lane, Chilwell,
Nottingham NG9 5JN
0602 225049

Complete production service in video and film, broadcast standard and U-matic standard television. Also offer advisory and consultancy services.

Service Training Ltd
34 The Square, Kenilworth,
Warwickshire CV8 1EB
0926 512421/2

Complete programme production in video, film, slide/filmstrip, for sales and/or training purposes. Individual services of writers, photographers and artists available. Expertise in technical, engineering and commercial fields.

Slide Centre Ltd
143 Chatham Road, London
SW11 6SR
01-223 3457

Slide duplicating (long or short runs), slide shooting location photography, audio production, duplication and distribution. Specialists in 'package deal' slide set production.

Slide Graphics
36 Southwell Road, London
SE5 9PG
01-733 6119

Produce colour transparencies from black-and-white artwork. Audio visual production.

Slides Visual Communications Ltd
Murray House, Vandon Street, London SW1
01-222 1194

Complete audio visual production in slide/tapes. Everything from script and story board. Production studio and location photography. Mixing and programming facilities. Consultancy on slide/tape presentations and the associated hardware. Specialise in conferences, exhibitions, PR, and product launches.

Slidewise (AV) Ltd
Unit One, 158 High Street, Bushey, Hertfordshire WD2 3LN
01-950 5959

Slide duplicating, artwork production and artwork copying. Slide/tape programme production (scripting, visualising, photography, sound recording and pulsing). Production of OHP transparencies.

Slidework (London) Ltd
42-44 Beak Street, London
W1R 3DA
01-439 4357

Sound and Vision Communications
24 Redan Place, London W2 4SA
01-229 4406

Slide/tape production house. Facilities available for hire also. Editing facilities, 3 production rooms, sound transfer bay, theatre containing 3 channel sound and AVL Show Pro 5C with 15 carousel projectors.

Sound Communications (Publishers) Ltd
Field House, Wellington Road, Dewsbury, West Yorkshire
WF13 1HF
0924 451717

Duplication of cassettes and open reel tapes. Artwork design and print service. Studio facilities for voice-overs. Supply blank cassettes.

Soundcraft Network Video Ltd
17 Whittle Road, Ferndown Industrial Estate, Wimborne, Dorset BH21 7RL
0202 875466

Video tape production in colour and black-and-white, and soundtracks (for video tape or slide/tape).

Sound Developments Ltd
7 Chalcot Road, London
NW1 8LH
01-586 1271

Audio visual mainly for conferences including slides, video and sound recording.

Sound Education (Publishers)
2 Hastings Court, Collingham, Wetherby, West Yorkshire
LS22 5JL
0937 72698

Produce and market tailor-made audio visual programmes (slide/tape, films, tape and work-book) on any subject, especially management and training.

South London Video
137 Railton Road, London SE24 0LY
01-733 0323

Any video production from broadcast to industrial promotion tapes.

Sparfax Television
Cheney Court
Ditteridge
Wiltshire SN14 9QF
0225 743487/742204

Programme consultants for use of video in industry. Producers of industrial TV programmes for management communications, staff training, direct selling exhibitions and conferences. All facilities in-house. Promotional and public relations material for broadcast and industrial quality transmission.

Spectrum Communications Ltd
183/185 Askew Road, London
W12 9AX
01-749 3061

Film, video, audio-visual, photographic and publishing.

Spectrum Video and Film Production Services Ltd
14 Charlotte Mews, Tottenham Street, London W1
01-631 0608

Film and video production company using Ikegami 3 tube cameras and Sony U-matic video high band and low band editing units.

Spoken Image
The Loft, 5 Cambridge Street, Manchester M15 JG
061-236 7522

Design and production in slide/tape (complete service). 6000 sq ft of space, rostrum cameras, photographic unit, sound recording studio.

SSK Visuals Ltd
The Studios, Newton Terrace Lane, Glasgow G3 7PB
041-226 3851

Complete audio visual production service. Facilities include graphics studio, 8 track mixing desk and sound booth. Programming using Electrosonic single and multi-screen, and AVL Sho-Pro 5 computerised multiple projection system, rostrum camera 35/46mm Forox registration for slide production and colour processing.

Stanmore Video Services Ltd
91-93 High Street, Edgware, Middlesex HA8 7DB
01-951 0466

8mm and 16mm film transferred to video. 35mm slides transferred to video. Video tape duplication. Hire facilities for video equipment and studio including 2 Ikegami 3 tube cameras, cox mixer and sound and lighting.

Beryl Stevens Associates
1st Floor, 12-13 Henrietta Street, London WC2
01-836 0437

Complete audio visual production service specialising in employee communications. Film, video, slide/tape. Supporting literature: wallcharts, booklets, etc. Part production service also offered.

Stewart Films
Cygnet Guild Communications Ltd
2 Orchard Road, Malvern, Worcestershire WR14 3DA
068 45 4975

Film production including location and animation. Complete service.

Sudbury Conference Hall
Sudbury House
15 Newgate Street
London EC1
01-248 1202

Preview theatre with facility for 16 and 35mm film projection up to cinemascope size. Twin screen 35mm slide projection using Electrosonic equipment. Sound recording facilities on ¼in tape. Also well-known as a conference hall with the above facilities together with public address and refreshment facilities.

SVC Slides
Murray House, Vandon Street, London SW1
01-222 1194

Slide preparation and duplication, and full audio visual production.

Swift Film Productions
1 Wool Road, London
SW20 0HN
01-946 2040

Complete in-house production company specialising in industrial productions and documentaries. 16mm films, audio visual programmes, filmstrips, 16mm films loops. Slide/tape to film transfer. Foreign versions. Adviser on computer generated films.

Synchrovisual Ltd
Steven's Mead, The Green, Chalfont St Giles, Buckinghamshire HP8 4QA
024 07 5732

Slide/tape and filmstrip/tape production from basic conception. Complete or part service. Full service in foreign language versions including oriental languages.

Talkiestrips (part of the Cygnet Guild group)
Guild House, Upper St Martin's Lane, London WC2H 9EL
01-240 1073

Production house for all audio visual formats of slide/tape and filmstrip. Transfer either of these to video or 16mm film.

Talking Pictures Productions Ltd
41 Paradise Walk, London
SW3 4JW
01-351 1151

Audio and slide/tape productions with hire facilities.

Teamwork AV
34 Cricklewood Broadway, London NW2 3ET
01-450 0266

All in-house facilities. Shoot colour slides from own black-and-white artwork. Audio visual production including concept and script, photography, mixing, dubbing, in-house AVL equipment including Sho-Pro 5b. Full conference design and staging. Equipment hire.

Theatre Projects Services
10-16 Mercer Street, London WC2
01-240 5411

Slide projection, PMTs. Full sound hire, any AVL projection equipment, programming facilities, viewing theatre, full registration camera.

Thorn-EMI Ltd
Thorn House, Upper St Martin's Lane, London WC2H 9ED
01-836 2444

Specialise in sponsored public relations and prestige film and video production: medical, technical, scientific, training, sales, safety. Music and internal film and video programming.

3 Arrows Film and TV Productions
17 Convent Walk, Sheffield
S3 7RX
0742 737821

A complete production company providing 16mm and 35mm film and colour video production for prestige, training and sales, pilot and TV commercials. 16mm double head and video preview theatre available for hire. Also servicing for professional equipment.

TPR Audio Visual Ltd
2a Prospect Crescent, Harrogate, North Yorkshire HG1 1RH
0423 502518

Audio visual slide/tape production company. Entire programme including scripting and full photographic facilities. Sound studio with pulsing facilities.

Treo Video Ltd
1-7 Boundary Row, London
SE1 8HP
01-633 9494

Outside broadcast (OB) facilities, electronic field production (EFP) facilities. Studio and post-production facilities on BCN one-inch broadcast standard. Agents for hardware.

Ralph Tuck Promotions Ltd
14 Stradbroke Road, Southwold, Suffolk IP18 6LQ
0502 723571

Video training films to client specifications. Complete or part service.

Unison Films Ltd
41 Carshalton Road, Sutton, Surrey SM1 4LG
01-643 7277

United Motion Pictures
Boston House, 36/38 Fitzroy Square, London W1P 5LL
01-580 1171

Film production and post-production facilities.

Video Arts Ltd
2nd Floor, Dumbarton House, 68 Oxford Street, London W1N 9LA
01-637 7288

Producers and distributors of training films and video cassettes for industry and education (both to specification and pre-recorded).

Video Inclusive Ltd
49-51 Norwood Road, Herne Hill, London SE24 9AA
01-674 5580

Light-weight video productions. Specialise in role playing and appraisal situations. Single camera work. Full lighting and sound back-up. All post production services.

Videotron Ltd
443 Cranbrook Road, Gants Hill, Ilford, Essex 1G2 6EW
01-554 7617

Small budget production work in video. Sell and hire video and audio visual equipment.

Videoview
68-70 Wardour Street
London W1V 3HP
01-437 1333

Video software; language series. Distributor for Video Media.

Viscom Group
25-27 Farringdon Road, London
EC1
01-404 5041

Viscom offers film, video and slide/tape production facilities. Viscom Slides offers the full range of duplication of 35mm and 46mm transparencies and other related services. Viscom produces complete audio visual communications programmes in internal communications and training applications.

Visnews Ltd
Cumberland Avenue, London
NW10 7EH
01-965 7733

Production of documentary films, video programmes and other audio visual packages. Complete in-house facilities. Also facilities for recording, editing and distributing all types of video material.

Vistech
74 Brodrick Road, London
SW17 7DY
01-672 1751

Audio visual production in education, training and careers. Complete service.

Visual Aids Service
12-16 North Street, Carshalton, Surrey
01-669 4900

Production of simple instructional films. Colour conversion slides, slide presentations.

The Visual Connection (TVC) Ltd
50 Glebe Place, London
SW3 5JE
01-352 5177/9500

Audio visual presentations from single projector to multi-screen including scripts, story boards, visual production (eg photographic artwork, etc). Conversions from audio visual to video and films. Sound tracks, effects, voice-overs, etc.

Visual Mar-Com Systems Ltd
Mar-Com House, Thames Road, Strand-on-the-Green, Chiswick, London W4 3PP
01-995 8345

OHP transparencies, 35mm slides and filmstrips to specification.

Visual Presentations
92 York Street, London W1
01-724 2461/2

Presentations in 35mm slides, slide/tapes and Superslides (artwork included). Duplication. Facilities to produce artwork for exhibition stands.

Visual Resources Ltd
25 High Street, Alton, Hampshire
GU34 1AW
0420 83058

Photographic library and range of visual products and services to aid AV producers, advertising agencies, designers, printers etc.

Wall to Wall Stage Shows Ltd
First Floor, 92 Bristol Street, Birmingham
021-622 6463

Promotions, concert promotions, recording, management and publishing.

Warren Recordings
59 Hendale Avenue, London
NW4 4LP
01-203 0306

Duplication of cassettes. Supply blank cassettes tailored to specified length. Studio and mobile recording facilities for hire.

Western Sound Visual Ltd
Bristol Video Centre,
198a Cheltenham Road, Bristol
BS6 5QZ
0272 47823

Video production to specifications, scripting, shooting and production. 16mm commag and comopt. Telecine facility. Studio facilities available for hire.

West Films
159 Whiteladies Road, Bristol
BS8 2RF
0272 35924

Chris Whitmore Associates Ltd
Chestnut Road, Glenfield,
Leicester LE3 8DB
0533 873545

Graphics and audio visual slide/tape production. Also complete conference organising. Production of slides (stills, animated sequences, etc).

Woodmansterne Ltd
Holywell Industrial Estate,
Watford, Hertfordshire
WD1 8RD
92 28236

Tape/slide, slide and filmstrip productions.

Worldwide Pictures
21-25 St Anne's Court, London
W1V 3AW
01-434 1121

Produce tailor-made programmes for industry, commerce and government. Work in all media: film, video tape, video disc and slide/tape. Operate 3 sound recording theatres (also for hire). Animation company (Worldwide Animation Company) produce graphics, titling and animation.

WSTV
159-163 Great Portland Street
London W1N 5FD
01-323 1144

Specialist in video tape production. Industrial and commercial.

Diana Wyllie Ltd
1 Park Road, Baker Street,
London NW1 6XP
01-723 7333

Produce filmstrip, slide/cassette, filmstrip/cassette programmes for education and training.

Yorkshire Film Co Ltd
Robeck House, Victoria Road,
Huddersfield, West Yorkshire
HD1 3RB
0484 25335

Complete production service in 16mm films (documentary style) including scripting. 16mm rock and roll dubbing facilities for hire. Complete conferences/exhibitions services.

Zaar International Cinema and Television Programmes Ltd
339 Clifton Drive South,
St Anne's-on-Sea, Lancashire
0253 721053

16mm and 35mm film and video production facilities (including scripting and translating). Own dubbing suite with 8 channels of synchronous playback to picture with rock and roll. 2 editing suites (also for hire).

Zoom Television Ltd
Unit 11, Cowley Mills Trading Estate, Longbridge Way,
Uxbridge, Middlesex
01-895 7981

Make industrial documentaries and internal communication programmes for industry and commerce. Produce on high-band BVU Sony format and on normal low-band Sony U-matic format. Complete service including scripting, technical and post production.

TRAINING AND INFORMATION

TRAINING

HIGHER DEGREES

University College of Wales,
Aberystwyth
Department of Education
Cambrian Chambers
Cambrian Place
Aberystwyth SY23 1NU
Wales
0970 3111

MEd(Wales) in Educational Technology

University of Bath
Claverton Down
Bath BA2 7AY
0225 61244

MEd in Educational Technology; PhD in Educational Technology

University of Birmingham
Department of Educational Psychology
Faculty of Education
Ring Road North
PO Box 363
Birmingham B15 2TT
021-472 1301

BPhil (Ed) in Educational Technology

University College, Cardiff
Registrar, University College
PO Box 78
Cardiff, South Wales
0222 44211

MEd in Educational Technology

University of Glasgow
Institute of Educational Studies
Glasgow, Scotland
041-339 8855 ext 7452

MEd in Curriculum Development and Educational Technology

University of Hull
Institute of Education
173 Cottingham Road
Hull HU5 2EH
0482 46311

MA in Communication Learning and Individual Differences in Education

Padgate College of Higher Education, Fearnhead
Warrington
Cheshire WA2 0DB
0925 814343

BEd (Hons) University of Manchester in Audio Visual Communications

University of Surrey
Institute for Educational Technology
Guildford, Surrey GU2 5XH
0483 71281 ext 402

MPhil or PhD in Educational Technology; MSc in Teaching and Learning in Higher Education

University of York
Heslington, York Y01 5DD
0904 59861

MPhil or DPhil in Educational Studies; MA in Applied Educational Studies

POSTGRADUATE DIPLOMAS

Note: A first degree is not always necessary in order to be admitted to the diploma courses listed below. Some colleges will take those with other professional qualifications or on the basis of relevant experience.

College of Wales, Aberystwyth
Department of Education
Cambrian Chambers
Cambrian Place
Aberystwyth SY23 1NU
0970 3111

College Diploma in Educational Technology

University of Birmingham
Department of Educational Psychology
Faculty of Education
Ring Road North, PO Box 363,
Birmingham B15 2TT
021-472 1301

Diploma in Educational Technology

City University
Centre for Educational Technology
London EC1V 4PB
01-253 4399

Jointly with:
Hatfield Polytechnic
Education Department
Hatfield
Hertfordshire AL10 9AB
30 68100

and

University of Surrey
Institute for Educational Technology
Guildford, Surrey GU2 5XH
0483 71281

Diploma in Teaching and Learning in Higher Education

Coventry Technical College
Butts, Coventry CB1 SG9
0203 57221

Diploma in Educational Technology

Dundee College of Education
Gardyne Road, Broughty Ferry,
Dundee DD5 1NY
0382 453433

Advanced Diploma in Educational Technology; CNAA Postgraduate Diploma in Educational Technology; Diploma in Educational Technology

Dundee College of Technology
Bell Street, Dundee DD1 1HG
0382 27225

College Diploma in Educational Technology

Garnett College
Downshire House
Roehampton Lane
London SW15 4HR
01-789 6533

University of London Diploma in Educational Technology

Hatfield Polytechnic
PO Box 109, College Lane,
Hatfield, Hertfordshire
A110 9AB
30 68100

CNAA Postgraduate Diploma in Applied Educational Studies

Jordanhill College of Education
76 Southbrae Drive
Glasgow G13 1PP
041-959 1232

CNAA Postgraduate Diploma in Educational Technology

University of London
Institute of Education
20 Bedford Way, London
WC1H 0AL
01-636 1500

Diploma in Teaching and Course Development

Middlesex Polytechnic
Bounds Green Road
London N11 2NQ
01-368 1299

Polytechnic Diploma in Educational Technology

Newcastle-upon-Tyne Polytechnic
Ellison Place
Newcastle-upon-Tyne NE1 8ST
0632 26002

CNAA Diploma in Educational Technology

Plymouth College of Art and Design
Drake's Circus, Plymouth, Devon
0752 21312

Diploma in Audio Visual Techniques for Education

Plymouth Polytechnic
Polytechnic Hoe Centre
Notte Street, Plymouth, Devon
0752 21312

MSc Diploma in Educational Technology; CNAA Postgraduate Diploma in Educational Technology; Polytechnic Diploma in Educational Technology

South Devon Technical College
Newton Road, Torquay, S Devon
0803 35711

Diploma in Educational Technology

South Thames College
Department of Educational Resources
Wandsworth High Street
London SW18 2PP
01-870 2241

College Diploma in Learning Resources

University of Surrey
Institute for Educational Technology
Guildford, Surrey GU2 5XH
0483 71281

Postgraduate Diploma in Teaching and Learning in Higher Education

W R Tuson College
St Vincents Road, Fulwood,
Preston PR2 4UR
0772 716511

Diploma in Educational Technology

University of York
Heslington, Yorkshire YO1 5DD
0904 59861

Diploma in Educational Broadcasting

OTHER COURSES IN EDUCATIONAL TECHNOLOGY AND ASSOCIATED SUBJECTS

Note: A number of local education departments run courses in presentation and equipment handling, either as part of a broader course or with reference to the efficient use of a specific piece of equipment. The majority of these courses tend to be run on an *ad hoc* basis in response to suggestions from teachers in the area. It is always worth contacting your local authority to make suggestions for such courses and to find out what courses are available.

Courses are also run by certain manufacturers and suppliers of audio visual aids, but these too are seldom arranged on a regular basis and are often concerned with the use of specific pieces of equipment. Information about many of these courses is given in the regular audio visual journals.

Association for Educational and Training Technology
BLAT Centre for Health and Medical Education
BMA House, Tavistock Square
London WC1
01-388 7976

Seminars and conferences

Audio Visual Programmes Ltd
Studio AVP, 82 Clifton Hill
London NW8 0JT
01-624 9111

Short courses in communication and presentation for management.

Bell and Howell Audio Visual Ltd
Alperton House
Bridgewater Road, Wembley
Middlesex
01-902 8812

One-day courses for projectionists (16mm projector).

British Association for Commercial and Industrial Education (BACIE)
16 Park Crescent, London
W1N 4AP
01-636 5351

Courses in training analysis, design and techniques; training technology; management and interpersonal skills; communication skills.

Cartrefle College
North East Wales Institute
Wrexham, Clwyd LL13 9HL
0978 51782

Advanced Diploma in Educational Technology

Coventry Technical College
Butts, Coventry
West Midlands CV1 3GD
0203 57221

CGLI 730 Educational Technology; CGLI 731 Educational Technology (two years)

CTVC
Hillside, Merry Hill Road, Bushey
Hertfordshire WD2 1DR
01-950 4426

Short courses in radio and television presentation techniques

Cybernetics Teaching Systems Ltd
89 Park Lane, Castle Donington
Derby DE7 2JG
0332 810639

Criterion referenced instruction

Eastleigh College of Further Education
Chestnut Avenue, Eastleigh
Hampshire
0703 614444

Short courses for (a) office practice teachers; (b) workshop instructors

Enlightenment Audio Visual Productions
St Catherine's, Botesdale, Diss
Norfolk IP22 1BP
037989 434

Tailor-made seminars for managers on effective presentation

Guardian Business Services Ltd
21 John Street
London WC1N 2BL
01-278 2332

Public courses in the use of audio visual aids and equipment.

Huddersfield Polytechnic
Hollybank Road, Lindley
Huddersfield HD3 3BP
0484 25611

Courses by special arrangement for industrial trainers. Educational technology, micro-teaching, interview techniques, counselling and guidance and general audio visual production.

Huntingdon Technical College
California Road, Huntingdon
Cambridgeshire PE18 7BL
0480 52346

Short course in instructional techniques. Further Education Teachers Course.

The Industrial Society
PO Box 1BQ, Robert Hyde House
48 Bryanston Square
London W1H 1BQ
01-262 2401

Seminars for training managers.

The Institute of Training and Development
5 Baring Road, Beaconsfield
Buckinghamshire HP9 2NX
04946 3994

Seminars in training and diploma in training management.

Lancashire College for Adult Education
Southport Road, Chorley
PR7 1NB
02572 76719

Courses in educational technology; courses in TV production; courses in radio production; resource-based language courses; seminars and conferences connected with the media and resource-based learning.

Learning Systems Group
Middlesex Polytechnic at Bounds Green
Bounds Green Road, London N11
01-368 1299

Various short courses in aspects of educational technology including programmed instruction, simulation and gaming, design of learning resources, use of video in training.

National Audio-Visual Aids Centre
Film Library, Paxton Place
London SE27 9SR
01-670 4247/8/9

Short intensive courses in all aspects of media production and educational technology. Tailor-made courses either at the Centre or in-house. Consultancy service.

National Health Service
Learning Resources Unit
Sheffield City Polytechnic
55 Broomgrove Road
Sheffield S10 2NA
0742 661862

Courses for nurse educators in the production and use of resources in teaching and learning.

Oldham College of Technology
Rochdale Road, Oldham
Lancashire OL9 6AA
061-624 5214

CGLI 224 Electronics Servicing course for mechanics with fourth-year option in audio equipment.

Plymouth Polytechnic
Drake Circus, Plymouth P14 8AA
0752 21312

Various specialist short courses in educational technology and educational television.

RECALL Training Consultants
13 Wisteria Road
London SE13 5HW
01-852 5775

Workshops on educational technology, in-company, instructor training; working languages: English, German, French, Italian.

Saville Audio Visual Ltd
Salisbury Road
York YO2 4YW
0904 52011

Organises Video Schools which run regular courses at venues in the north of England. Courses include a one-day 'Introduction to Video' course and a two-day course on 'Practical Video' for the more advanced user.

Sheffield City Polytechnic
Department of Education Services
30 Collegiate Crescent
Sheffield S10 2BP
0742 20911

Workshop courses in the design and development of learning resources and active training methods, programmed instruction, simulations and games, role-play, case studies, slide/tape packages.

Slough College of Higher Education
Department of Development
Training Technology Division
Wellington Street
Slough, Berkshire
75 34585

Courses in applications of educational technology to industrial training.

South Devon Technical College
Newton Road, Torquay
Devon TQ2 5BY
0803 35711

CGLI 731 Educational Technology (two years). Also evening modular course in Audio Visual Resources for Teachers.

Southgate Technical College
High Street, London N14 6BS
01-886 6521

Audio visual aid courses for teachers. CGLI 736 Audio Visual Aid Course for Technicians (two years).

South Thames College
Department of Educational Resources
Wandsworth High Street
London SW18 2PP
01-788 4442

Certificate in media resources (12 weeks, each term). Short courses covering different aspects of educational technology including a projectionist course and an introduction to film-making and photography.

The Suffolk College of Higher and Further Education
Rope Walk, Ipswich
Suffolk IP4 KT
0284 5656

CGLI 731 Educational Technology (two years).

Television Training Centre
23 Grosvenor Street, London W1
01-629 5069

Diploma in Television Studies, Television Direction and Production.

3M Visual Workshops
3M House, PO Box 1
Bracknell, Berkshire RG12 1JH
0344 26726

Transparency Workshops

University Teaching Methods Unit
55 Gordon Square
London WC1H 0NT

Courses and seminars for teachers in higher education.

PROFESSIONAL ORGANISATIONS

Aslib
Audio Visual Group
3 Belgrave Square
London SW1X 8PL
01-235 5050

Association for Educational and Training Technology (formerly APLET)
BLAT Centre for Health and Medical Education
BMA House, Tavistock Square
London WC1H 9JP
01-388 7976

Association of Teachers of Mathematics
Market Street Chambers
Nelson, Lancashire BB9 70N
0282 63360

Association of Video Dealers
Greystones
Brighton Road
Godalming
Surrey GU7 1PL
04868 23429

Audio Visual Association
139 Greenwich High Road
London SE10
01-853 5153

British Association for Commercial and Industrial Education
16 Park Crescent
London W1N 4AP
01-636 5351

British Association for Early Childhood Education
Montgomery Hall
Kennington Oval
London SE11 5SW
01-582 8744

British Computer Society
13 Mansfield Street
London W1N 0BP
01-637 0471

British Council Media Department
Tavistock House South
Tavistock Square
London WC1H 9LL
01-388 2494

British Educational Equipment Association
Sunley House
10 Gunthorpe Street
London E1 7RW
01-247 9320

British Federation of Film Societies
81 Dean Street
London W1V 6AA
01-437 4355

British Film Institute
127-133 Charing Cross Road
London WC2H 0EA
01-437 4355

British Industrial and Scientific Film Association
26 D'Arblay Street
London W1V 3FH
01-439 8441

British Institute of Recorded Sound
29 Exhibition Road
London SW7
01-589 6603/4

British Kinematograph, Sound and Television Society
110-112 Victoria House
Vernon Place
London WC1B 4DJ
01-242 8400

British National Film Catalogue
British Film Institute
127-133 Charing Cross Road
London WC2
01-437 4355

British Universities Film Council
81 Dean Street
London W1V 6AA
01-734 3687

Centre for Information on Language Teaching and Research
20 Carlton House Terrace
London SW1Y 5AP
01-839 2626

Children's Film Foundation Ltd
6-10 Great Portland Street
London W1N 6JN
01-580 4796

Conservation Trust Resource Centre
c/o The George Palmer School
Northumberland Avenue
Reading
0734 868442

Council for Educational Technology for the United Kingdom
3 Devonshire Street
London W1N 2BA
01-580 7553/4

Educational Films of Scotland
Dowanhill
74 Victoria Crescent Road
Glasgow G12 9JN
041-334 9314

Educational Foundation for Visual Aids (EFVA)
Film Library
Paxton Place, London SE27
01-670 4247

Educational Television Association
(formerly NECCTA)
86 Micklegate
Yorkshire YO1 1JZ
0904 29701

Films of Scotland
32a Rutland Square
Edinburgh EH1 2BW
031-229 3456

Group for Educational Services in Museums
c/o Mr G Carter
Countryside Educational Trust
Beaulieu Manor
Hampshire SO4 7ZN
0590 612345

Historical Association
59a Kennington Park Road
London SE11 4JH
01-735 2974

The Industrial Council for Educational and Training Technology Ltd
Leicester House
8 Leicester Street
London WC2H 7BN
01-437 0678

Institute of Amateur Cinematographers
63 Woodfield Lane
Ashtead, Surrey
27 76358

London International Film School
24 Shelton Street
London WC2
01-240 0168

Media Studies Association
Leigh College, Railway Road
Leigh WN7 4HA
0942 671271

Midlands Universities and Polytechnics Committee on Educational Technology
Audio Visual Aids Unit
University of Technology
Loughborough
0509 63171 ext 461

Museums Association
34 Bloomsbury Way
London WC1A 2SS
01-404 4767

National Audio-Visual Aids Centre (NAVAC)
Film Library
Paxton Place, London SE27
01-670 4247

National Committee for Audio-Visual Aids in Education (NCAVAE)
Film Library
Paxton Place, London SE27
01-670 4247

National Film School
Beaconsfield Film Studios
Station Road, Beaconsfield
Buckinghamshire
04946 71234

National Institute of Adult Education (NIAE)
19b De Montfort Street
Leicester LE1 7GE
0533 551451

National Inter-College Committee on Educational Technology (NICCET)
Craigie College of Education
Ayr KA8 0SR
0292 60321

National Reprographic Centre for Documentation
Hatfield Polytechnic
Endymion Road Annexe
Hatfield
Hertfordshire AL10 8AU
30 66144

The Royal Photographic Society
The Octagon
Milsom Street, Bath, Avon
0225 62841

Royal Television Society
Tavistock House East
Tavistock Square
London WC1H 9HR
01-387 1970

Scottish Association of Amateur Cinematographers
Dowanhill
74 Victoria Crescent Road
Glasgow G12 9JN
041-334 9314

Scottish Council for Educational Technology
Dowanhill
74 Victoria Crescent Road
Glasgow G12 9JN
041-334 9314

Scottish Educational Media Association
Dowanhill
74 Victoria Crescent Road
Glasgow G12 9JN
041-334 9314

Society for Education Through Art
Bath Academy of Art
Corsham, Wiltshire SN13 0DB
0249 712571

Society for Education in Film and Television
29 Old Compton Street
London W1V 5PL
01-734 5455/3211

Society for Academic Gaming & Simulation in Education and Training (SAGSET)
Centre for Extension Studies
University of Technology
Loughborough
Leicestershire LE11 3BT
0509 63171

Standing Conference on Educational Development Services in Polytechnics
c/o Dr P Griffin (Secretary)
Learning Resources Unit
Middlesex Polytechnic
Bounds Green Road
London N11 2NQ
01-368 1299

LIST OF FILM LIBRARIES

BBC Enterprises Film Hire
Woodston House
Oundle Road
Peterborough PE2 9PZ

British Petroleum Film Library
15 Beaconsfield Road
London NW10 2LE

Central Film Library
Government Building
Bromyard Avenue
Acton, London W3 7JB

Concord Films Council Ltd
201 Felixstowe Road
Ipswich, Suffolk IP3 9BJ

Concordia Films
Viking Way
Bar Hill Village
Cambridge CB3 8EL

CTVC
Hillside
Merry Hill Road
Bushey, Watford WD2 1DR

Walt Disney Educational Media Company
83 Pall Mall
London SW1

Education and Television Films Ltd
247A Upper Street
London N1 1RU

Eothen Films Ltd
EMI Film Studios
Shenley Road
Borehamwood
Hertfordshire WD6 1JE

Gateway Educational Media
Waverley Road
Yate, Bristol BS17 5RD

Guild Sound & Vision Ltd
Woodston House
85-129 Oundle Road
Peterborough PE2 9PZ

National Audio Visual Aids Library
Film Library
Paxton Place, London SE27

National Film Board of Canada
1 Grosvenor Square
London W1X 0AB

Open University Film Library Department
Guild Sound & Vision Ltd
Woodston House
85 129 Oundle Road
Peterborough PE2 9PZ

Edward Patterson Associates Ltd
68 Copers Cope Road
Beckenham, Kent

Random Film Library
25 The Burroughs
London NW4 4AT

Rank Aldis
Rank Audio Visual
PO Box 70
Great West Road
Brentford
Middlesex TW8 9HR

Scottish Central Film Library
Dowanhill
74 Victoria Crescent Road
Glasgow G12 9JN

Shell Film Library
25 The Burroughs
London NW4 4AT

TFI (Training Films International)
St Mary's Street
Whitchurch
Shropshire

Town and Country Productions
21 Cheyne Row
Chelsea
London SW3 5HP

Transport, Travel and Electricity Film Library
Melbury House
Melbury Terrace
London NW1 6LP

Video Arts
Dumbarton House
68 Oxford Street
London W1N 9LA

Viscom
Audio Visual Library
Park Hall Trading Estate
London SE21 8EZ

Welsh Office Film Library
Oxford House
Hills Street
Cardiff CF1 2XG

PROFESSIONAL AND TRADE JOURNALS

Audio Visual
Maclarens, monthly

Audio Visual Directory & Buyers' Guide
Maclarens, annually

Audiovisual Librarian, The
Aslib, quarterly

BACIE Journal
BACIE, monthly

The BKSTS Journal
BKSTS, monthly (10 issues)

British Journal of Educational Technology
Longman Group, thrice yearly

Broadcast
Broadcast, weekly

BUFC Newsletter
BUFC, thrice yearly

Camera Club Journal
Camera Club, monthly

Camerawork
Half Moon Photography Workshop, bi-monthly

Educational Broadcasting International
British Council, quarterly

Educational Computing
Richard Hease, 10 issues per year

Educational Media International
ICEM/NCAVAE, quarterly

Film
British Federation of Film Societies, monthly

Film and Television Technician
ACTAT, monthly

Films and Filming
Hanson Books, monthly
Greater London Arts Association, bi-annual

Journal of Education for Teaching
Methuen, 3 times per year

Journal of Educational Television and other media
Educational Television Association, quarterly

Media, Culture and Society
Academic Press, quarterly

Media International
Alain Charles, monthly

Monthly Film Bulletin
BFI, monthly

Pictorial Education
Evans, monthly

Pictorial Education special
Evans, bi-monthly

Practical Education
Educational Institute of Design, Craft and Technology/EIDCT, quarterly

Programmed Learning and Educational Technology
AETT/Kogan Page, quarterly

Records and Recording
Hansom Books, monthly

Screen
SEFT, quarterly

Screen Digest and CTV Report
Screen Digest/Nord Media Ltd, monthly

Screen Education
SEFT, quarterly

Screen International
SEFT, quarterly

Sight and Sound
BFI, quarterly

Simulation/Games for Learning
Society for Academic Gaming and Simulation in Education and Training/Kogan Page, quarterly

Training Digest
John Chittock, monthly

Video
Link House Magazines, monthly

Videogram
Nord Media Ltd, Bi-monthly

Video Review
IPC, monthly

Video Today
Modmags, monthly

Video World
Galaxy Publications Ltd, monthly

Video Yearbook
Link House Magazines, annually

Viewdata and TV User
IPC Electrical and Electronics Press, quarterly

EQUIPMENT HIRING AGENCIES

Code for equipment hiring agencies:

F = Film equipment
FS = Filmstrip equipment
OHP = Overhead projector equipment
S/T = Slide/tape equipment
V = Video equipment

Applied Audio Visual Ltd
Bridges Mews, Bridges Road
Wimbledon, London SW19 1EL
01-540 0651
S/T V

Associated Visual Products
Prospect House
14-20 Prospect Place
Welwyn Garden City
Hertfordshire AL6 9EN
043-871 7931/2
F OHP

Audio Visual Distributors
29 Patrick Square
Edinburgh EH8 9EY
031-667 9231
F FS OHP S/T

Audio Visual Equipment Ltd (AVE Ltd)
73 Surbiton Road, Kingston
Surrey KT1 2HG
01-549 3464
F FS OHP S/T V

Audio Visual Leisure
Unit 2
West Parade Industrial Estate
Halifax, West Yorkshire HX1 2TF
0422 58600
F OHP V

Audio Visual Material Supply Co
AVM House
1 Alexandra Road
Farnborough, Hampshire
0252 40721/3
F S/T

Audio Visual Services
J Hudson Ltd
45 Stroud Road
Gloucester GL1 5AQ
0452 35181
F FS OHP S/T V

Audio Visual Techniques
86 Preston Road
Brighton, Sussex BN1 6AE
0273 561126
S/T

Automation Facilities Ltd
Blakes Road, Wargrave
Berkshire RG10 8AW
073 522 3012
V

AVM Video
Unit 26
Northfield Industrial Estate
Beresford Avenue, Wembley
Middlesex
01-902 9707
V

Bellard Electronics Ltd
Video House
115 Long Lane, Upton
Chester CH2 1JS
0244 43671
V

Benedict-Lindand Ltd
1-2 Newman Passage
Newman Street
London W1G 3HG
01-637 7809
F V

Better Sound Ltd
33 Endell Street
London WC2
01-836 0033
F

Boness Audio-Visual Services Ltd
2-3 Clifton Terrace
Finsbury Park
London N4
01-272 5606
F

Brabury Electronics Ltd
119A Loverock Road
Reading RG3 1NS
0734 52434
V

Briana Electronics Ltd
White Lodge
East Hanningfield Road
Sandon, Chelmsford
Essex CM2 7TQ
0245 71145
F V

British Films Ltd
Carlyle House
235 Vauxhall Bridge Road
London SW1V 1EJ
828 7965
F OHP

Cadbury/Schweppes Ltd
Bournville
Birmingham B30 2LU
021-458 2000
S/T

Cameron C W Ltd
Burnfield Road
Giffnock
Glasgow G46 7TH
041-633 0077
V

Cine Europe Ltd
245 Old Marylebone Road
London NW1 5QT
01-402 8385
F

Cinephoto Equipment Ltd
Cinephoto House
17 The Crescent
Salford M5 4PF
061-736 6221/2
F FS OHP

Cinevideo Ltd
245 Old Marylebone Road
London NW1 5QT
01-724 1363
F V

Concord Films Council Ltd
201 Felixstowe Road
Ipswich, Suffolk IP3 9BJ
0473 76012
V

Copywrite Machines Ltd
44-46 Sillwood Street
Brighton, Sussex BN1 2PS
0273 720175
OHP S/T

CPD, Consolidated Photovisions Distribution Ltd
No 3 Evans Road
Peel Green, Eccles,
Manchester M30 7LY
061-707 4638

Currivans Photographic Service
53 Keeper Road
Dublin 12
Eire
0001 752307
F OHP

Dale Hire Ltd
16 Bermondsey Trading Estate
Rotherhithe New Road
London SE16
01-231 2050
S/T V

R E Davis
36 St Botolph's Street
Colchester, Essex CO2 7EA
0206 73444
F FS OHP S/T

Decron Audio-Visual Services
12 Castle Avenue
Datchet, Berkshire SL3 9BA
75 43424
V

Drake AV Video Ltd
212 Whitchurch Road
Cardiff CF4 3NB
South Glamorgan
0222 24502/3
F FS OHP S/T V

Edco Reed International Productions Ltd
Broadway Chambers
London W6
01-741 1921
S/T

Edinburgh Film Productions
Nine Mile Burn
Penicuik
Midlothian EH26 9LX
0968 72131
F

Edric Audio Visual Ltd
34/36 Oak End Way
Gerrards Cross
Buckinghamshire SL9 8BR
02813 84646 or 0272 555119
(British Branch); 061-773 7711
(Manchester Branch)
F FS OHP S/T V

EFVA
Sales Department
Film Library
Paxton Place, London SE27 9SR
01-670 4247/8/9
F OHP

EFVA
Video and Sound Studio Facilities
Film Library
Paxton Place, London SE27 9SR
01-670 4247/8/9
V

Electronic Picture House
191 Wardour Street
London W1
01-439 9701
V

Electrosonic Ltd
815 Woolwich Road
London SE7 8LT
01-855 1101
F S/T

George Elliott and Sons Ltd
Ajax House
Hertford Road, Barking, Essex
01-591 5599
F FS OHP S/T

EOS Electronics AV Ltd
Western Square, Barry
South Glamorgan CF6 7TY
0446 741212
V

Erricks AV and Video
Fotosonic House
Rawson Square, Bradford
West Yorkshire BD1 3JR
0274 22972
F OHP S/T

Evans R T and Sons
43-49 New Street
Stevenston KA20 3HE
Ayrshire
0294 64154
F

Evershed Power-Optics Ltd
Bridge Wharf, Bridge Road
Chertsey, Surrey KT16 8LJ
093 28 61181
F

Fisher Audio Visual
Audio Visual Photographic Equipment
724 Borough Road
Birkenhead
Merseyside L42 9JE
051-608 4848
F OHP S/T V

Focus Films (Leeds) Ltd
281 Meanwood Road
Leeds LS7 2JR
0532 625267
F

Fotokine (Reading) Ltd
357 Oxford Road
Reading, Berkshire RG3 1AZ
0734 54746
F S/T

Fraser-Peacock
94 High Street
Wimbledon Village
London SW19 5EG
01-947 7551
F FS S/T

G D P – Samuelson Sight and Sound
19 Calthorpe Road
Edgbaston
Birmingham B15 1RP
021-455 0818
F FS OHP S/T

Gordon Audio Visual Ltd
28-30 Market Place
Oxford Circus
London W1N 8PH
01-580 9191
F FS OHP S/T

Gordon Hammonds Audio Visual Co Ltd
97 Alma Road
Southampton, Hampshire
0703 35575
F S/T

Gray Audio Visual
Kingsley Works
Grange Road
London NW10 2RE
01-451 0247
F OHP S/T

Guild Sound and Vision Ltd
Woodston House
Oundle Road
Peterborough PE2 2PZ
F OHP S/T V

Gye Handley Associates
9 Dean Street
London W1
01-439 8705
S/T

Hadland, John (PI) Ltd
Newhouse Laboratories
Newhouse Road
Bovington
Hemel Hempstead
Hertfordshire
0442 832525
F

Hammond AV and Video Services
60 Queens Road
Watford
Hertfordshire WD1 2QN
93 39733/7
F FS OHP S/T V

Hargreaves
204-206 Warbreck Moor
Aintree
Liverpool L9 0HZ
051-525 6458/9
F FS OHP S/T V

Hire Service Shops Ltd
Warenne House
31 London Road, Reigate
Surrey RH2 9PZ
74 49441
F

Hocken, H Ltd
33 Station Road
Redhill, Surrey RH1 1QP
91 62776
F FS OHP S/T V

Hedges Wright (AV) Ltd
40 Victoria Road
Swindon, Wiltshire SN1 5QZ
0793 31032
F S/T

Holden (Photographics) Ltd
49 Fishergate
Preston PR1 8AQ
0772 58038
F OHP S/T V

Holiday Bros (Audio and Video) Ltd
172 Finney Lane
Heald Green, Cheadle
Manchester
061-437 0538/9
V

Hudson and Carter Ltd
3 Attenborough Lane
Chilwell
Nottinghamshire N69 5JN
V

Hughes TV and Audio
17-21 White Lion Street
Norwich NR2 1QA
0603 60935
V

ICP Films Ltd
The Studio Centre
Ackhurst Road
Chorley, Lancashire
02572 66411
F FS S/T V

Independent Television
ITN House
48 Wells Street
London W1P 4DE
01-637 2424
F V

Istead Ltd
38 The Tything
Worcester
0905 29713/4/5
F FS OHP S/T V

J B Presentations Ltd
15 Brackenbury Road
London W6 0BE
01-749 6036
F FS S/T

Kadek Vision Ltd
Shepperton Studio Centre
Studios Road, Shepperton
Middlesex TW17 0QD
09328 66941
F FS OHP S/T

Keith Johnson Photographic
Ramillies House
Ramillies Street, London W1
01-734 5684
F OHP S/T

Kent Video
16 Sherman Road
Bromley, Kent
01-464 9979
V

John King Films Ltd
Film House
71 East Street
Brighton, Sussex BN1 1NZ
0273 202671
F FS OHP S/T V

Lane End Productions Ltd
9 Meard Street
London W1V 3QH
01-734 0072
F V

Lauries (Film Services) Ltd
Audio Visual Centre
Robeck House, Victoria Road
Lockwood, Huddersfield
Yorkshire HD1 3RB
0484 37122
F FS S/T V

Livesey Audio Visual
95 Princes Avenue, Hull
North Humberside HU5 3QR
0482 43453
F FS OHP S/T V

Lomax Ltd
8 Exchange Street
St Anne's Square
Manchester M2 7HL
061-832 6167
F FS S/T

London Video Ltd
Marvic House
Bishops Road
London SW6 7AD
01-385 8443/4

Martak Northern
Unit 12
Knutsford Industrial Estate
Park Gate Lane, Knutsford
Cheshire
0565 54655
FS S/T

Martak Southern Ltd
86-94 High Street
Alton
Hampshire GU34 1EN
0420 88011
FS S/T

McKenna and Brown Ltd
Video and AV Dept
190 Linthorpe Road
Middlesborough
Cleveland TS1 2JT
0642 246708
F OHP S/T V

Media Facilities (Scotland) Ltd
3 Jamaica Street
Edinburgh EH3 6HH
031-225 1910/226 2960
F FS S/T V

Mediatech
Woodside Place, Alperton
Wembley, Middlesex
01-903 7279
F S/T

Merseyside Audio Consultants
PO Box 4
Liverpool L23 3DT
051-924 8010
OHP S/T

Middlesex Film Productions Ltd
Shepperton Studio Centre
Studios Road, Shepperton
Middlesex TW17 0QD
093 28 60221
FS

Midland Audio Visual
The Rear
202 New Road, Rubery
Birmingham B45 9BA
021-453 3141
F OHP S/T

Midland Film Productions Ltd
Harrison Road
Birmingham B24 9AB
021-373 0450
F S/T

Midlands Video Systems
Video Centre
3a Attenborough Lane
Nottingham NG9 5JN
0602 252521
V

Molland, Tom Ltd
Audio Visual Centre
110 Cornwall Street
Plymouth, Devon PL1 1NN
0752 669282
F S/T V

MST Presentations
Spencer Court
7 Chalcot Road
London NW1 8LH
01-722 0088/9
S/T

Multi-Vision Audio Visual Ltd
Wellington House
Duke Street
Manchester M3 4NF
061-834 9994
F S/T

Myriad AV Sales
10-11 Great Newport Street
London WC2
01-240 1941
S/T V

National Sound Reproducers Ltd
3 Parr Road, Stanmore
Middlesex HA7 1PZ
01-951 0222/5
F OHP S/T V

John Norman (Audio Visual) Ltd
St Helier Station
Green Lane, Morden, Surrey
01-648 7916
F S/T

Northern Sound Services Ltd
Broad Chare, Quayside
Newcastle-upon-Tyne NE1 3EB
0632 26304
F FS OHP S/T

Opsis Ltd
83 Prospect Road, Southborough
Tunbridge Wells
Kent TN4 0EE
0892 42708
F FS OHP V

Pelling and Cross Ltd
104 Baker Street
London W1M 2AR
01-487 5411
F FS OHP S/T

Penrose Cine (Dollands Photographic Ltd)
69 Streatham Hill
London SW2 4TZ
01-674 5602
F OHP

Photoco AV
427-431 London Road
Sheffield S2 4HJ
0742 53351
F OHP S/T V

Prater Audio Visual Ltd
139 Greenwich High Road
London SE10
01-858 5715
FS S/T

Projection and Display Services Ltd
Newton Works, Stanlake Mews
Stanlake Villas
London W12 7HS
01-749 2201/3
F

QAV Presentations Ltd
7 Guildford Road
Woking, Surrey
048 62 70204
F FS S/T V

Rainbow Video
48 Lower Addiscombe
Lower Addiscombe Road
Croydon, Surrey CR0 6AA
01-680 1940
V

Rands Video Ltd
Video House
10 Hibel Road
Macclesfield
Cheshire SK10 2AB
0625 26248
V

Rank Audio Visual
PO Box 70
Great West Road
Brentford
Middlesex TW8 9HR
01-568 9222
F

REW Ltd
10-12 High Street
Colliers Wood
London SW19
01-540 9684/5/6
V

Research Recordings Ltd
5-7 Clarendon Terrace
Edgware Road, Maida Vale
London W9 1BZ
01-289 0123
V

Robian Visual Services
43 King Street
Stanford-le Hope
Essex SS17 0HJ
03756 43777
F OHP

Roundel Productions Ltd
51 Loudon Road
London NW8 0DL
01-624 6080
S/T

Samuelson Sight and Sound
303-315 Cricklewood Broadway
Edgware Road
London NW2 6PQ
01-452 8090
F V

SB Modules Ltd
159 Great Portland Street
London W1N 5FD
01-323 1144
S/T

Sarner Audio Visual
32 Woodstock Grove
Shepherds Bush, London W12
01-743 1288
F FS OHP S/T V

Saville Audio Visual Ltd
Salisbury Road, York YO2 4YW
0904 37700
S/T

SAV Presentations Ltd
Ridgeway House
132 Tanners Drive, Blakelands
Milton Keynes MK14 5BR
0908 612586
FS OHP S/T V

SAV Studios Ltd
34 Cricklewood Broadway
London NW2 3ET
01-278 7893
S/T

Scottish Council for Educational Technology
Dowanhill
74 Victoria Crescent Road
Glasgow G12 9JN
041-334 9314
F FS OHP S/T V

Sheffield Photo Co Ltd
6 Norfolk Row, Fargate
Sheffield S1 1SN
0742 22241
F OHP V

Simmon Sound and Vision
28a Manor Road
Bradford, Yorkshire
0274 307763
F

Smart Sound and Vision
13 Hawkins Road, Earlson
Coventry, West Midlands
0203 77449
F FS S/T

Solent Audio Visual Ltd
228 London Road
Portsmouth, Hampshire
0705 62091
OHP S/T

Sound and Vision Communications
24 Redan Place
London W2 4SA
01-229 4406
F FS S/T V

Sound and Visual Products Ltd
25 Nicholson Square
Edinburgh EH8 8BX
031-667 9231
F FS OHP S/T

Soundcraft Network Video
17 Whittle Road
Ferndown Industrial Estate
Wimborne, Dorset
0202 875466/7
F FS S/T V

Sound Developments Ltd
7 Chalcot Road
London NW1 8LH
01-586 1271
S/T

South London Video
137 Railton Road
London SE24 0LT
01-733 0323
V

Specialist Audio Visual
127 Trafalgar Road, Greenwich
London SE10 9TX
01-858 9160
F OHP S/T

SSK Visuals Ltd
The Studios
Newton Terrace Lane
Glasgow G37 PB3
041-226 3851
FS S/T

Stanmore Video Services Ltd
91-93 High Street
Edgware, Middlesex HA8 7DB
01-951 0466
V

Studio 45 Audio Visual Centre Ltd
45 Templar Avenue, The Hill
Coventry CV4 9BQ
0203 461341
F FS OHP S/T V

Studio 99 Video Ltd
249 Cricklewood Broadway
Edgware Road
London NW2 6NX
01-450 1313
F V

Talking Pictures Productions Ltd
41 Paradise Walk
London SW3 4JW
01-351 1151
FS S/T

Teamwork AV
34 Cricklewood Broadway
London NW2 3ET
01-450 0266
S/T

Tele Tape Video Hire Department
12 Golden Square
London W1R 3AF
01-434 3311
V

Television International Operations Ltd
9-11 Windmill Street
London W1P 1HF
01-637 2477
V

Terry-More Photographic
49 George Street
Luton, Bedfordshire LU1 2AQ
0582 23391/2
F FS OHP

Theatre Projects Services
10-16 Mercer Street
London WC2
01-240 5411
S/T

The Visual Connection (TVC) Ltd
50 Glebe Place
London SW3 5JE
01-352 5177/9500
S/T

TPR Audio Visual
2a Prospect Crescent
Harrogate
North Yorkshire HG1 1RH
0423 502518
FS OHP S/T

Trilion Video Ltd
Video House
36-44 Brewer Street
London W1
01-439 4177
V

Ralph Tuck Promotions Ltd
14 Stradbroke Road
Southwold
Suffolk IP18 6LQ
0502 723571
V

Turners (Audio Visual) Ltd
7-15 Pink Lane
Newcastle-upon-Tyne NE1 5HT
0632 25391
F FS OHP S/T

Tyne Video Ltd
South Shore Road
East Gateshead Industrial Estate
Gateshead
Tyne and Wear NE8 3AE
0632 775627
V

United Film Services
13 King Street
Leicester LE1 6RN
0533 542777
F FS OHP S/T

Video Cassette Recorders Ltd
112 Long Acre
Covent Garden
London WC2 9NT
01-836 5717
V

Video Inclusive Ltd
49-51 Norwood Road
Herne Hill, London SE24 9AA
01-674 5580
V

Videoscan Ltd
Oxford Road Mill
Macclesfield
Cheshire SK11 8HP
0625 612000
V

Video South
101 Eden Vale Road
Westbury, Wiltshire
0373 823140
F V

Video Transfer Services
134 Ealing Road, Wembley
Middlesex
01-903 2008
F V

Visual Mar-Com Systems Ltd
Mar-Com House
Thames Road
Strand-on-the-Green
London W4 3PP
01-995 8345
F FS OHP S/T V

Western Cine Services
30 Alphington Road
Exeter, Devon EX2 8HN
0392 56651
OHP S/T V

Western Sound Visual Ltd
Bristol Video Centre
198a Cheltenham Road
Bristol BS6 5QZ
0272 47823
V

Yorkshire Film Co Ltd
Robeck House
Victoria Road
Huddersfield
West Yorkshire HD1 3RB
0484 25335
F S/T

Zoom Television Ltd
Unit 11
Cowley Mills Trading Estate
Longbridge Way
Uxbridge, Middlesex
01-895 7981
V

SELECTIVE BIBLIOGRAPHY

GENERAL AUDIO VISUAL EDUCATION

The Audio Visual
By Jack Dove
André Deutsch, 1975

Audio-Visual Aids and Techniques in Managerial and Supervisory Training
By R P Rigg
Hamish Hamilton, 1969

Audio-Visual Aids: An Introduction
National Committee for Audio-Visual Aids in Education
NCAVAE, 1973

Audio-Visual Handbook
By Ralph Cable
University of London, 1973

Audio-Visual Processes in Education: Selected Readings from ***A-V Communication Review***
AECT, 1971

A-V Instruction – Media and Methods
By James *et al*
McGraw-Hill, 1973

Bibliography of Selected Titles on School Broadcasting
School Broadcasting Council, 1977

Bibliography of Museum and Art Gallery Publications and Audio-Visual Aids in Great Britain and Ireland
Chadwick-Healey, 1978

The Box in the Corner
By Gwen Dunn
Macmillan, 1977

CELPIS – A second list of some audio-visual and other materials made by colleges and departments of education and teachers' centres
CET, 1976

Children and Television
Edited by Ray Brown
Collier Macmillan, 1977

Children in the Picture
By Frank Blackwell
CET, 1975

The Colour Factor Set in Infant and Junior Schools
NCAVAE, Occasional Paper No 17

Communication and Learning
By L S Powell
Pitman, 1972

Communication and Schools
By C W Bendin
Pergamon, 1970

The Concept of Educational Television
By Kenneth Richmond
Weidenfeld and Nicolson, 1970

The Conditions of Learning
By Robert M Gagné
Holt, Rinehart and Winston, 1977

Contributions to an Educational Technology
By I K Davies and J Hartley
Butterworths, 1972

Contributions to an Educational Technology Vol 2
Edited by I K Davies and J Hartley
Kogan Page, 1978

Current British Research on Mass Media and Mass Communication: Register of Ongoing and Recently Completed Research
By Connie Ellis
University of Leicester, 1978

Degree of Difference: A Study of the First Year's Intake of Students to the Open University of the United Kingdom
By Naomi E McIntosh *et al*
Society for Research into Higher Education, 1976

Designing Instructional Text
By J Hartley
Kogan Page, 1978

The Development of Cognitive Processes
By V Hamilton and M D Vernon
Academic Press, 1977

Developments in Communication
By D Gardner
CET, 1974

Educational Technology: A Contribution to the Improvement of Education
CET, 1978

Educational Technology and Team Teaching: The Way Ahead
NCAVAE, Occasional Paper No 15, 1968

Educational Technology in Curriculum Development
By Derek Rowntree
Harper and Row, 1974

Educational Television and Radio
BBC, 1966

Educational Value of Non-Educational Television
By M Scarborough
IBA Fellowship Report, 1974

Effective Presentation
By Antony Jay
Management Publications, 1972

Encyclopaedia of Educational Media, Communications and Technology
Edited by D Unwin and R McAleese
Macmillan, 1978

Evaluation of Instruction Materials
By W J Webster
AECT, 1976

The Evaluation of Schools Programmes
By G Kemelfield
IBA Fellowship Report, 1972

Exploring Material with Young Children
By Roy Sparkes
Batsford, 1975

Filmstrip and Slide Projectors in Teaching and Training
By Brydon Lamb
NCAVAE, 1971

Formative Evaluation of Educational Television Programmes
Seminar and Conference Report edited by Tony Bates and Margaret Gallagher
CET, 1979

Fool's Lantern or Aladdin's Lamp? The Use of Educational Television with Slow Learning and Handicapped Children
By Roy Edwards
IBA Fellowship Report, 1974

Fundamentals of Teaching with Audio-Visual Aids
By W H Erikson and D H Curl
Macmillan, 1972

Future of Educational Telecommunications: A Planning Study
By G W Tressel, D P Buckelew, J T Suchy and P L Brown
Lexington Books, 1975
UK distributor Teakfield Ltd

Graphics Simplified
By A J MacGregor
University of Toronto Press, 1979

A Guide to the Overhead Projector
By L S Powell
BACIE, 1974

A Guide to the Use of Visual Aids
By L S Powell
BACIE, 1973

HELPIS 5 – A catalogue of some audio-visual materials made by institutions of higher education
BUFC, 1978

Instructional Media and the Individual Learner
By E U Heidt
Kogan Page, 1978

Instructional Technology: Its Nature and Use
By W A Wittich and C F Schuller
Harper and Row, 1973

The Intelligent Eye
By R L Gregory
Weidenfeld and Nicolson, 1970

International Dictionary of Education
By G Terry Page, J B Thomas and A R Marshall
Kogan Page, 1979

Into Print
By John Gough
Batsford 1979

Keller Plan in the Classroom
Edited by D W Daley and S M Robertson
SCET, 1978

Learning to Teach Practical Skills
By Ian Winfield
Kogan Page, 1979

Lights Please! Using Projectors in the Classroom
By Robert Leggat
NCAVAE, 1972

Microcircuits, Society and Education
By W Gosling
CET Occasional Paper No 8, 1978

Modern Developments in Audiology
By James Jerger
Academic Press, 1973

Modern Teaching Aids
By N J Atkinson
Applied Science Publications, 1975

The 1978 Multi-Media International Year-book
Multi-Media International, US, 1978

New Methods and Media in Further Education
By Brydon Lamb
NCAVAE, 1970

Planning for Educational Mass Media
By A Hancock
Longman, 1977

The Practice of Educational Psychology
By Chazan, Moore, Williams and Wright
Longman, 1974

Primary School Innovation and Technology
By A Robinson and J Embling
CET, 1975

Producing for Educational Mass Media
By A Hancock
Longman, 1976

Psychology in the Use of Audio-Visual Aids in Primary Education
By G Mialaret
UNESCO, 1966

The Psychology of Visual Perception
By R Haber and M Hershinson
Reinhart and Winston, 1974

Realities of Teaching: Explorations with Video-Tape
By R S Adams and B J Biddle
Holt, Rinehart and Winston, 1970

A Select Bibliography of Educational Technology
By S Stagg and M Eraut
CET, 1975

Selecting Media for Learning
AECT, 1976

The Selection and Use of Instructional Media
By A J Romiszowski
Kogan Page, 1974

Simplified Techniques for Preparing Visual Material
By E Minor
McGraw-Hill, 1973

'Showing Off' or Display Techniques for the Teacher
By Robert Leggat
NCAVAE, 1974

Status and Trends of Distance Education
By B Holmberg
Kogan Page, 1981

A Survey of British Research in Audio-Visual Aids, 1945-1971
By Helen Coppen
NCAVAE, 1971

A Survey of British Research in Audio-Visual Aids: Supplement 1, 1972-1973
By Susie Rodwell
NCAVAE, 1973

A Survey of British Research in Audio-Visual Aids: Supplement 2, 1974
By Susie Rodwell
NCAVAE, 1975

A Survey of British Research in Audio-Visual Aids: Supplement 3, 1975
By Susie Rodwell
NCAVAE, 1976

A Survey of British Research in Audio-Visual Aids: Supplement 4, 1976
By Susie Rodwell
NCAVAE, 1977

A Survey of British Research in Audio-Visual Aids: Supplement 5, 1977
By Susie Rodwell
NCAVAE, 1978

Teaching and Learning Aids
By J Cummins
NIAE, 1972

Teaching and Training: Techniques for Instructors
By H R Mills
Macmillan, 1972

Teaching at a Distance
By E G Wedell and H D Perraton
NIAE, 1968

Teaching by Projection
By R S Judd
Focal Press, 1963

Teaching Materials for Disadvantaged Children
By R Gulliford and P Widlake
Schools Council Curriculum Bulletin 4, Evans/Methuen, 1974

Team Teaching in Britain
By John Freeman
Ward Lock, 1969

Techniques for Producing Visual Instructional Media
By Minor and Frye
McGraw-Hill, 1977

Television and Children
By Michael J A Howe
New University Education, 1977

Television and the Pre-School Child
By Harvey Lesser
Academic Press, 1977

Television Productions for Education
By P Coombes and J Tiffin
Focal Press, 1978

The Training of Teachers in Educational Technology
National Committee for Audio-Visual Aids in Education
NCAVAE, 1971

The Use of Resources
By J Hanson
Allen and Unwin, 1975

Using Audio-Visual Materials in Education
By James S Kinder
Van Nostrand Reinhold, 1965

Videocassettes in Education and Training
Edited by J Leedham and A J Romiszowski
Kogan Page, 1975

Visual Aids and Photography in Education
By M J Langford
Focal Press, 1975

Visual Aids: Their Construction and Use
By G C Weaver and F W Bollinger
Van Nostrand Reinhold, 1949

Visual Awareness
By F Palmer
Batsford, 1972

Visual Information Processing
Edited by W G Chase
Academic Press, 1973

Wallsheets, Choosing, Using and Making
By Helen Coppen
NCAVAE, 1971

RESOURCES AND RESOURCE CENTRES

Administrating Educational Media
By J W Brown and K Norberg
McGraw-Hill, 1972

Allocation and Management of Resources in Schools
By E Briault
CET Occasional Paper No 6, 1974

Area Resource Centre: An Experiment
By Emmeline Garnett
Edward Arnold, 1972

A Bibliographic System for Non-Book Media
By Antony Croghan
Coburgh Publications, 1979

A Challenge to Librarians
By R Fothergill
CET Working Paper No 4, 1971

Copyright Agreements between Employers and Staff in Education
By Geoffrey Crabb
CET, 1978

Copyright Clearance: A Practical Guide
By Geoffrey Crabb
CET, 1977

Curriculum Development: A Comparative Study
By P H Taylor and M Johnson
NFER, 1974

Criteria for Planning the College and University Learning Resources Center
I Merrill and H Drob
AECT, 1977

The Design of Learning Spaces
By P Smith
CET, 1974

Developing Multi-Media Libraries
By W B Hicks and A Tillin
Binker, 1970

Information in the School Library: An Introduction to Non-Book Materials
By M R Shifrin
Bingley, 1973

Directory of Information Sources and Advisory Services in Educational Technology
By O Fairfax
CET, 1974

Inside a Curriculum Project
By M D Shipman
Methuen, 1974

Libraries of the Future
By J C R Liklider
MIT Press, 1965

National Resources for Education: A Handbook for Teachers
SCET, 1975

Non-Book Materials: Cataloguing Rules
CET, 1973

Non-Book Materials in Libraries: A Practical Guide
By R Fothergill and I Butchart
Clive Bingley/Linnet Books, 1978

Non-Book Materials: Their Bibliographic Control
By L A Gilbert and J W Wright
CET Working Paper No 6, 1971

Non-Book Media in Junior Schools: A Handbook of Practical Advice
By Peter Jones
School Library Association, 1978

Not by Books Alone
Edited by C Waite and B Colebourn
School Library Association, 1975

The Organisation of Audio-Visual Resources for Learning in a Local Education Authority
National Committee for Audio-Visual Aids in Education
NCAVAE

Organising Resources
By N Beswick
Heinemann, 1975

Planning and Operating Media Centers
AECT, 1976

Producing Guides to Local Resources
By R Ireland
CET, 1979

Producing Lists of Learning Materials
By Barbara Beswick
CET, 1979

Resource-Based Learning in Post Compulsory Education
By Pat Noble
Kogan Page, 1980

Resource Centres
By M L Holder and R Mitson
Methuen, 1974

Resource Organisation in Primary Schools
By Cecilia Gordon
CET, 1979

Resource Organisation in Secondary Schools: Report of an Investigation
By P Thornbury, I Gillespie and G Wilkinson
CET, 1979

Resources and Resource Centres
By J Walton and J Ruck
Ward Lock, 1975

A Resources Centre is a State of Mind
Scottish Educational Film Association, 1973

Resources Centres in Colleges of Education
By R Fothergill
CET Working Paper No 10, 1973

Resources for Learning
By L C Taylor
Penguin, 1972

Resources in Schools
By R P A Edwards
Evans, 1973

School Building Design and Audio-Visual Resources
National Committee for Audio-Visual Aids in Education
NCAVAE, 1975

School Library Resources Centres: Recommended Standards
The Library Association, 1973

School Resource Centres: Schools Council Working Paper 43
By Norman W Beswick
Evans/Methuen, 1972

The Setting up of a Resources Centre: 1, Basic Ideas
By Adam H Malcolm
Scottish Educational Film Association, 1974

The Setting up of a Resources Centre: 2, Planning and Staffing
By R N Tucker
Scottish Educational Film Association, 1976

The Setting up of a Resources Centre: 3, Retrieval Systems
By Adam H Malcolm
Scottish Educational Film Association, 1976

Standards for Cataloguing Non-Print Materials
By A Tillin and W Quinly
AECT, 1976

MEDIA STUDIES

Bad News, Volume 1
By Glasgow University Media Group
Routledge and Kegan Paul, 1976

Big Business and the Mass Media
By R Rubin
Lexington Books, 1978.
UK distribution by Teakfield Ltd

Capital and Culture: German Cinema 1933-45
By Julian Petley
BFI, 1979

A Change of Tack: Making 'The Shadow Line'
By B Sulik
BFI, 1976

Characteristics of Local Media Audiences
By Ray Brown
Saxon House, 1978

Children in Front of the Small Screen
By Grant Noble
Constable, 1975

Communications
By Raymond Williams
Penguin, 1976

Dangling Conversations: Book 1, The Image of the Media
By Brian Winston
Davis-Poynter, 1973

Dangling Conversations: Book 2, Hardware Software
By Brian Winston
Davis-Poynter, 1974

Don Siegel
By A Lovell
BFI, 1975

Dynamics of Television
By Jon Baggaley and Steve Duck
Saxon House, 1976

Elements of Films
By Lee Bobker
Harcourt, Brace and World, 1976

The Film Idea
By S Solomon
Harcourt, Brace and World, 1976

Film Language, A Semiotics of the Cinema
By C Metz
Oxford University Press, 1974

Film Study
By Frank Manchel
Tantivy Press

Film Teaching
By Hall, Knight, Hunt and Lovell
BFI, 1968

The Fleet Street Disaster
By Graham Cleverley
Constable, 1976

Football on Television
Edited by E Buscombe
BFI, 1975

Four Arguments for the Elimination of Television
By Jerry Mander
Wm Morrow and Co Inc, 1978

Grierson on Documentary
Edited by Forsyth Hardy
Faber, 1979

Guerrilla Television
By Michael Shamberg
Holt, Rinehart and Winston, 1971

Hazell: The Making of a Television Series
By Manual Alvarado and Ed Buscombe
BFI/Latimer, 1978

History of the Cinema: From its Origins to 1970
By E Rhode
Penguin, 1978

Is This Your Life? Images of Women in the Media
By Stott, King *et al*
Virago, 1977

Journalists at Work
By Jeremy Tunstall
Constable, 1971

Journey into Journalism
By Arnold Wesker
Writers and Readers Publishing Cooperative, 1978

Labour Power in the British Film Industry
By M Chanan
BFI, 1976

Light Entertainment
By R Dyer
BFI, 1973

The Long View
By Basil Wright
Paladin, 1976

Making 'Legend of the Werewolf'
By E Buscombe
BFI, 1976

The Manufacture of News
Edited by Stanley Cohen and Jock Young
Constable, 1973

Making the Papers: The Access of Resource-Poor Groups to the Metropolitan Press
By E H Goldenburg
Lexington Books, 1975.
UK distributor Teakfield Ltd

The Mass Media
By P Golding
Longman, 1974

Mass Media and Cultural Relationships
By A Piepe, S Crouch and M Emerson
Saxon House, 1978

Mass Media and Mass Man
By Alan Casty
Holt, Rinehart and Winston, 1973

Mass Media in the Classroom
By Brian Firth
Macmillan, 1968

The Media are American
By Jeremy Tunstall
Constable, 1977

The Media Men
By John Morrison
Batsford, 1978

Media Sociology: A Reader
Edited by Jeremy Tunstall
Constable, 1970

Message Dimension of Television News
By R S Frank
Lexington Books, 1973.
UK distributor Teakfield Ltd

Newspapers as Organisations
By Larz Engwall
Saxon House, 1978

Pictures on a Page
By Harold Evans
Heinemann, 1978

The Political Impact of Mass Media
By Colin Seymour-Ure
Constable, 1974

Politics and Cinema
By Andrew Sarris
Columbia University Press, 1979

The Production of Political Television
By Michael Tracey
Routledge and Kegan Paul, 1978

Putting 'Reality' Together – BBC News
By Philip Schlesinger
Constable, 1978

Reading Television
By J Aske and J Hartley
Methuen, 1978

Screen Reader 1
Society for Education in Film and Television, 1975

Semiology
By Paul Giraud
Routledge and Kegan Paul, 1975

Signs and Meaning in the Cinema
By Peter Wollen
BFI, 1974

The Silent Watchdog
By D Murphy
Constable, 1976

Structures of Television
By N Garnham
BFI, 1973

Studies in Documentary
By A Lovell and J Hillier
BFI, 1972

Talking about the Cinema
By J Kitses and A Kaplan
BFI, 1976

Techniques of Persuasion
By J A C Brown
Penguin, 1977

TV and Anti-Social Behaviour
By S Milgram and R L Shotland
Academic Press, 1973

Television, Censorship and the Law
By Colin R Munro
Saxon House, 1979

Television and People
By Brian Groombridge
Penguin, 1972

Television and the February 1974 Election
By T Pateman
BFI, 1974

Television and the Working Class
By A Piepe, M Emerson and J Lannon
Saxon House, 1975

Television Documentary Usage
By D Vaughan
BFI, 1976

Television Economics
By Bruce Owen, Jack Beebe and William Manning
Lexington Books, 1974.
UK distributor Teakfield Ltd

Television in Politics: Its Uses and Influences
By Jay G Blumler and Denis McQuail
Faber and Faber, 1968

Television News
By R Collins
BFI, 1976

Television Violence and the Adolescent Boy
By William Belson
Saxon House, 1978

Theories of Film
By A Tudor
BFI, 1975

Trade Unionism in Television
By Peter Seglow
Saxon House, 1978

Videology and Utopia
By A Willener, G Milliard and A Ganti
Routledge and Kegan Paul, 1976

Violence on the Screen
By A Gluckmann
BFI, 1971

Ways of Seeing
By John Berger
Penguin, 1972

Whose Media?
By Peter Lewis
Consumer Association, 1978

Whose News? Politics, the Press and the Third World
By Rosemary Righter
Times Books NY, 1978

World Communications
Prepared by UNESCO
Gower Press, 1975

RADIO, TELEVISION AND VIDEO

The Accessible Portapak Manual
By Michael Goldberg
Video Inn Vancouver, 1977.
UK distributor CATS

Adult Education and Television
Edited by Brian Groombridge
NIAE and UNESCO, 1966

Audio and Video Recording
By David Kirk
Faber and Faber, 1975

Basic TV Staging
By Gerald Millerson
Focal Press, 1974

Basic Video and Community Work
Edited by Andi Biren
Inter-Action, 1974

Beginners Guide to Television
By Gordon J King
Butterworths, 1975

Bristol Channel and Community Television
By P Lewis
IBA Fellowship Report, 1976

Cable Television
By J E Cunningham
Howard W Sams & Co/
The Bobbs Merill Co, USA, 1976

Closed Circuit Television Single Handed
By Tony Gibson
Pitman, 1972

Community: TV and Cable in Britain
By Peter M Lewis
BFI, 1978

Educational CCTV in Glasgow
NCAVAE, Occasional Paper No 14, 1968

TV Guidelines: Writing, Directing and Presenting
By Alistair J Wilson
Hutchinson, 1973

Effective TV Production
By Gerald Millerson
Focal Press, 1976

Everyday Television: Nationwide
By Charlotte Brunsdon and David Morley
BFI, 1978

An Experiment in Closed Circuit Television at Millfield School
By P Turner and C R M Atkinson
NCAVAE, 1972

Experiments in Television
By Tony Gibson
NCAVAE, 1968

Filming for Television
By A Englander and Paul Petzold
Focal Press, 1976

Five Video Environments
By M Hooykaas and E Stansfield
CATS, 1979

A Grammar of TV Production
By D Davis
Barrie and Rockliff, 1974

Local Television: Piped Dreams?
By Andrew Bibby and Cathy Denford with Jerry Cross
Redwing Press, 1979

The Making of a TV Series
By Philip Elliott
Constable, 1972

The Practice of ETV
By Tony Gibson
Hutchinson, 1970

Radio and Television
By Stuart Hood
David and Charles, 1976

Radio and Television – The Everyday Miracle
By Egon Larsen
Dent, 1976

Radio Broadcasting
Edited by R L Hilliard
Focal Press, 1975

Report of the Committee on the Future of Broadcasting
Chairman: Lord Annan
HMSO, 1977

The Selection of Radio Equipment for VHF Reception
By Tony Crocker
NCAVAE, 1972

Speak for Yourself
Local Radio Workshop, 1977

The Technique of Radio Production
By Robert McLeish
Focal Press, 1979

Television
By K Wicks
Macdonald Educational, 1972

Television: Behind the Screen
By Peter Fairley
ITV Publications, 1976

Television in the Service of a School
By Peter Turner
NCAVAE, 1973

TV Camera Operation
By Gerald Millerson
Focal Press, 1974

TV Lighting Methods
By Gerald Millerson
Focal Press, 1975

TV Sound Operations
By G Alkin
Focal Press, 1975

Understanding Sound and Video Recording
By Michael Overman
Lutterworth Press, 1977

Using Videotape
By Joseph F Robinson and P H Beards
Focal Press, 1976

The Video Distribution Handbook
Centre for Advanced Television Studies, 1978

Video in Community Development
By J Hopkins, C Evans, S Herman and J Kirk
CATS, 1973

Video in Scotland
By A Nealon and R Reynish
Studio 99 Video, 1977

Video: A User's Guide
By Norman Vigars *et al*
BISFA, 1977

The Video Yearbook 1979
Edited by Angus Robertson
Blandford Press, 1979

Videorecording: Record and Replay Systems
By G White
Butterworths, 1972

Videorecording for Schools
By C King
Evans, 1975

Videotape Recording
By Joseph Robinson
Focal Press, 2nd ed, 1979

VTR Workshop: Small Format Video
By L J Atienza
UNESCO, 1978.
UK distribution CATS

Work of the Television Journalist
By Robert Tyrell
Focal Press, 1972

Writing for Television and Radio
By Robert L Hilliard
Focal Press, 1976

TRAINING AND PROGRAMMED LEARNING

Academic Gaming and Simulation in Education and Training
By G I Gibbs and A Howe
Kogan Page, 1975

Aspects of Educational Technology IX: Educational Technology for Continuous Education
Kogan Page, 1976

Aspects of Educational Technology X: Individualised Learning
Kogan Page, 1977

Aspects of Educational Technology XI: The Spread of Educational Technology
Edited by P J Hills and J Gilbert
Kogan Page, 1978

Aspects of Educational Technology XII: Educational Technology in a Changing World
Edited by P Race and D Brook
Kogan Page, 1978

Aspects of Educational Technology XIII: Educational Technology 20 years on
Edited by G Terry Page and Quentin Whitlock
Kogan Page, 1979

Aspects of Educational Technology XIV: Educational Technology to the year 2000
Edited by Leo Evans and Roy Winterburn
Kogan Page, 1980

Aspects of Simulation and Gaming
Edited by Jacquetta Megarry
Kogan Page, 1977

Computer Assisted Learning in the UK: Some Case Studies
By R Hooper and I Toye
CET, 1975

Cybernetic Principles of Learning and Educational Design
By K U Smith
Holt, Rinehart and Winston, 1966

Designing Instructional Systems
By A J Romiszowski
Kogan Page, 1981

Education and Training for Film and Television, 2
By David Fisher and John Tasker
BKSTS, 1977

Games and Simulations in Science Education
H I Ellington, E Addinall and F Percival
Kogan Page, 1980

International Yearbook of Educational and Instructional Technology 1978/79
Edited by A Howe and A J Romiszowski
Kogan Page, 1978

Investigations of Microteaching
Edited by D McIntyre
Croom Helm, 1977

Learning and Programmed Instruction
By Taber *et al*
Addison-Wesley, 1965

Learning by Objectives
By A D Carroll, J Duggan and R Etchells
Hutchinson, 1979

A Manager's Guide to Coaching
By D Megginson and T Boydell
BACIE, 1979

Microcomputers in Secondary Education
Edited by J A M Howe and Peter Ross
Kogan Page, 1979

Microteaching
By Allen and Ryan
Addison-Wesley, 1969

Microteaching
By G Brown
Methuen, 1975

Microteaching in Higher Education: Development and Practice
By Elizabeth Perot
SRHE, 1977

Perspectives on Academic Gaming and Simulation 1 & 2: Communication, Computer Basis and Education
Edited by Jacquetta Megarry
Kogan Page, 1978

Perspectives on Academic Gaming and Simulation 3: Training and Professional Education
Edited by Ray McAleese
Kogan Page, 1978

Perspectives on Academic Gaming and Simulation 4: Human Factors in Games and Simulations
Edited by Jacquetta Megarry
Kogan Page, 1979

Perspectives on Academic Gaming and Simulation 5: Simulation and Gaming for the 1980s
Edited by Philip Race and David Brook
Kogan Page, 1980

Perspectives on Academic Gaming and Simulation 6: Simulation and Games: the Real and the Ideal
Edited by Betty Hollinshead and Mantz Yorke
Kogan Page, 1981

Programming Languages
Edited by F Glenuys
Academic Press, 1968

Readings in Computer Based Learning
Nick Rushby
Kogan Page, 1981

Selected Microteaching Papers
Edited by A J Trott
Kogan Page, 1977

Selection and Use of Instructional Media
A J Romiszowski
Kogan Page, 1978

The Self-Teaching Process in Higher Education
By P J Hills
Croom Helm, 1976

A Series of Case Histories of the use of Programmed Learning
NCAVAE, Occasional Paper No 16, 1968

Simulations – A Handbook for Teachers
Ken Jones
Kogan Page, 1980

A Systems Approach to Education and Training
Edited by A J Romiszowski
Kogan Page, 1970

PHOTOGRAPHY

Advanced Photography
By M J Langford
Focal Press, 1965

Basic Colour Photography
By Andreas Feininger
Thames and Hudson, 1972

Basic Photography
By Michael Langford
Focal Press, 1978

Better Photography
By Michael Langford
Focal Press, 1978

Beyond Photography
By Jack Tait
Focal Press, 1979

The Camera
By David Carey
Ladybird, 1970

Camera: A Victorian Eyewitness
By Gus MacDonald
Batsford, 1979

The Camera at War
By Jorge Lewinski
W H Allen, 1979

Camera Copying and Reproduction
By O R Croy
Focal Press, 1964

Commonsense Photography
By Leonard Gaunt
Focal Press, 1975

Complete Art of Printing and Enlarging
By O R Croy
Focal Press, 1976

A Concise History of Photography
By Gernsheim and Gernsheim
Thames and Hudson, 1965

Creative Photographic Printing Methods
By Harold C Woodhead
Focal Press, 1976

Creative Techniques in Nature Photography
By Arnold Wilson
Batsford, 1979

Croy's Camera Trickery
By O R Croy
Focal Press, 1977

Darkroom Techniques Vols 1 and 2
By Andreas Feininger
Thames and Hudson, 1974

Design by Photography
By O R Croy
Focal Press, 1971

Effects and Experiments in Photography
By Paul Petzold
Focal Press, 1973

Exploring Photography
By Bryne Campbell
BBC Publications, 1978

Focalguide to Cibachrome
By J Coote
Focal Press, 1978

Focalguide to Slides
By Graham Saxby
Focal Press, 1979

Focalguide to Slide-Tape
By Brian Duncalf
Focal Press, 1978

The Focalguide to Cameras
By Clyde Reynolds
Focal Press, 1977

The Focalguide to Colour Printing from Negatives and Slides
By Jack H Coote
Focal Press, 1976

The Focalguide to Effects and Tricks
By Gunter Spitzing
Focal Press, 1974

The Focalguide to Enlarging
By Gunter Spitzing
Focal Press, 1974

The Focalguide to Flash
By Gunter Spitzing
Focal Press, 1974

The Focalguide to Home Processing
By R E Jacobson
Focal Press, 1974

The Focalguide to Lighting
By Paul Petzold
Focal Press, 1977

The Focalguide to Lenses
By Leonard Gaunt
Focal Press, 1977

The Focalguide to Low Light Photography
By Paul Petzold
Focal Press, 1979

The Focal Encyclopaedia of Photography Volumes 1 and 2
Focal Press, 1977

Introducing Photograms
By P Brundet
Batsford, 1973

Introduction to Bird and Wildlife Photography
By J Marchington and A Clay
Faber, 1974

The Lens in Action
By Sidney F Ray
Focal Press, 1977

Manual of Professional Photography
By Philip Gotlop
Thames and Hudson, 1973

The Photographer's Handbook
By J Hedgecoe
Ebury Press, 1977

Photography
By Greenhill, Murray and Spence
Macdonald, 1977

Photography for Children
By John M Pickering
Batsford, 1976

Photography for Designers
By Julian Sheppard
Focal Press, 1976

Photography in School
By Robert Leggat
Fountain, 1975

Photography: Materials and Methods
By J Hedgecoe and M J Langford
Focal Press, 1975

Photography/Politics: 1
Edited by T Bennett, D Evans, S Gohl, Jo Spence and Lucy Williams
Photography Workshop, 1979

Photomontages of the Nazi Period
By John Heartfield
Gordon Fraser, 1977

Pocket Money Photography
By Christopher Wright
Pan, 1975

Professional Photography
By M J Langford
Focal Press, 1974

Simple Photography With and Without a Camera
By Peter Marmoy
Studio Vista, 1976

Slide Tape and Dual Projection
By R Beaumont-Cragg
Focal Press, 1975

Starting Photography
By M J Langford
Focal Press, 1976

Your Film and the Lab
By Bernard Happé
Focal Press, 1975

Focal Press publish many handbooks on the use of individual makes of camera.

REPROGRAPHY

Basics of Reprography
By Tyrell
Focal Press, 1972

Reprography for Librarians
By New
Bingley, 1975

Reprographic Management Handbook
By F C Crix
Business Books, 1975

Reprographic Principles Made Easy
By John Young
NCAVAE, 1975

AUDIO

The All in One Tape Recorder Book
By Joseph M Lloyd
Focal Press, 1975

Audio Cassettes: The User Medium
By Somanta Banerjee
UNESCO (France), 1978

The Audio Handbook
By Gordon J King
Butterworths, 1975

Beginners' Guide to Audio
By Sinclair
Butterworths

British Documentary Sound
By Alexander Ross
SCET, 1977

Choosing a Tape Recorder
NCAVAE, Occasional Paper No 2, 1971

Hi-Fi Yearbook 1979
Edited by Colin Sproxton
IPC Electrical Electronic Press, 1979

The Law and Your Tape Recorder
By A Phelon
Print and Press Services, 1966

Master Creative Tape Recording
By Gardner
Butterworths

Master Hi-Fi Loudspeakers and Enclosures
By D Berriman
Butterworths, 1979

Sound for Film and Television
BKSTS, 1972

Sound Effects on Tape
Print and Press Services, 1966

Sound from Microphone to Ear
By Walters *et al*
BKSTS, 1977

Sound in Educational Television
By Glyn Alkin
CEDO, 1968

Sound with Vision
By E G M Alkin
Butterworths/BBC, 1973

Technique of the Sound Studio
By A Nesbitt
Focal Press, 4th ed, 1979

Tape and Cine
Print and Press Services, 1973

The Tape Recorder in the Classroom
By John Weston
NCAVAE, 1973

The Tape Recorder in Scottish Schools Parts 1 and 2
Scottish Educational Film Association, Part 1 1968, Part 2 1971

Tape Recorders
By P Spring
Focal Press, 1967

Tape Recording from A to Z
By D Crawford
Kaye and Ward, 1975

Teaching With Tape
By J Graham Jones
Focal Press, 1972

The Use of Microphones
By Alec Nisbett
Focal Press, 1975

Wildlife Sound Recording
By John B Fisher
Pelham Books, 1977

SPECIAL SUBJECTS

Photography in Art Teaching
By Alan Kay
Batsford, 1973

Computer Assisted Learning in Science Education
Edited by G Beech
Pergamon Press, 1979

Computer Educational Aids and Resources for Teachers
British Computer Society Schools Committee
NCAVAE, 1977

Design Education in Schools
Edited by Bernard Aylward
Evans, 1973

Design Education: Problem Solving and Visual Experience
By Peter Green
Batsford, 1974

Film in English Teaching
Edited by R Knight
BFI, 1972

Handbook for Geography Teachers
Edited by M Long
Methuen, 1974

The Geographers' Vademecum
By J C Hancock and P F Whiteley
George Philip, 1977

The Geography Room and its Equipment
By R Cole
Geographical Association, 1968

Guide to Resources in Environmental Education
By Peter Berry
Conservation Trust, 1978

Source Book for Environmental Studies
By P S Berry
George Philip, 1975

The Uses of a Revolving Blackboard in Geography Teaching
By J A Bond
Geographical Association, 1968

Education Objectives for the Study of History: A Suggested Framework
By J Fines and J B Coltham
Historical Association, 1976

Handbook for History Teachers
By W H Burston and C W Green
Methuen, 1972

Resources for the Teaching of History in Secondary Schools
By Keith Randell
Historical Association, 1976

The Use of Film in History Teaching
By Nicholas Pronay
Historical Association, 1972

Dissemination of Innovation: Humanities Curriculum Project
Schools Council Working Paper No 5
By J Ruddock
Evans, 1976

Aspects of Language and Language Teaching
By W A Bennett
Cambridge University Press, 1968

The Audio-Visual Approach to Modern Language Teaching
Edited by P J Vernon
NCAVAE

Introduction to the Language Laboratory
By Professor John D Turner
University of London, 1965

The Language Laboratory and Language Learning
By Julian Dakin
Longman, 1973

The Language Laboratory and Modern Language Teaching
By E M Stack
Oxford University Press, 1971

The Language Laboratory in School
By Green, Oliver and Boyd
Methuen, 1975

Handbook for Modern Language Teachers
Edited by Alan W Hornsey
Methuen, 1975

The Visual Element in Language Teaching
By S Pit Corder
Longman, 1966

Visual Materials for the Language Teacher
By Andrew Wright
Longman, 1976

Using the Language Laboratory
Edited by Professor John D Turner
University of London, 1969

Child's Play Mathematics: The Use of Play Materials in School
By Baker, Wells, Stephens and Fielker
Evans, 1975

Mathematics and the Ten Year Old
Schools Council Working Paper 61
Evans/Methuen Educational, 1979

Mathematics by Visual Aids
BACIE, 1962

Aids to Reading
By J M Hughes
Evans, 1970

Reading Resources
Centre for Teaching of Reading
University of Reading, 1973

Reading Skills: A Systematic Approach
By Elizabeth Hunter
CET, 1977

Reading with Words in Colour
By C Gathegro
Educational Explorers, 1962

A Select List of Aids of Use in the Teaching of Recent History
By G R Brooks
Historical Association, 1971

Teachers' Centres with Provision for Reading
By E J Goodacre
University of Reading, 1975

Problems in Science Education, 9-13 Age Range
By F Aicken
IBA Fellowship Report, available free, 1973

The 16mm Film in School Science and Technology
Compiled by Brian Gee
Science Education Unit, College of St Mark and St John
Plymouth, 1979

Chalk Illustration: A Manual for Technical Teachers
By B Pringle
Pergamon, 1966

Resources in Special Education
Edited by Frank McKee
SCET Occasional Working Paper No 1, 1976

Teaching Materials for Disadvantaged Children
By J P Hilton
Methuen, 1974

FILM

Animated Film: Concepts, Methods, Uses
By Roy P Madsen
Focal Press, 1977

Animating Films Without a Camera
By J Bourgeois
Oak Tree Press, 1974.
UK distributor Ward Lock

The Animation Stand
By Zoran Perisic
Focal Press, 1976

Basic Motion-Picture Technology
By Bernard Happé
Focal Press, 1974

Cartoon Animation for Everyone
By Alan Cleave
Fountain, 1973

Cine Craft
By J D Beal
Focal Press, 1974

The Complete Book of Movie Making
By Tony Rose
Fountain Press, 1971

Computer Animation
Edited by John Halas
Focal Press, 1974

Creating Special Effects: For TV and Films
By Bernard Wilkie
Focal Press, 1976

Directing Motion Pictures
Edited by Terence Marner
Tantivy Press

Film Design
By Terence Marner
Tantivy Press

Film Making: A Practical Guide
By Linder
Prentice-Hall, 1976

Film Making in Schools
By Douglas Lowndes
Batsford, 1973

Film Making in Teaching
By Keith Kennedy
Batsford, 1972

Films on the Campus
By T Fench
Yoseloff

The Focalguide to Moviemaking
By Paul Petzold
Focal Press, 1976

Film Propaganda: Nazi Germany and Soviet Russia
By Richard Taylor
Croom Helm, 1979

The Focalguide to Shooting Animation
By Zoran Perisic
Focal Press, 1977

Getting Started in Film-making
By Lillian Schiff
Oak Tree Press, 1978.
UK distributor Ward Lock

Good Frames and Bad
By Susan M Markle
John Wiley, 1969

Grammar of the Film Language
By Daniel Arijon
Focal Press, 1976

A Guide to the 8mm Loop Film
By G H Powell and L S Powell
BACIE, 1967

Hints and Tips for the Movie Maker
By Robert Bateman
Fountain Press, 1968

How to Edit
By H Baddeley
Focal Press, 1968

How to Make Films at School
By J Beal
Focal Press, 1968

How to Plan Your 8mm Movies
By C V Willson
Focal Press, 1973

How to Shoot a Motion Picture
By Ivan Watson
Macmillan, 1979

How to Title
By L F Minter
Focal Press, 1962

Making Films in Super 8: A Handbook for Primary and Secondary School Teachers
By D R Huxley
Cambridge University Press, 1978

Manual of Film Making
By B Calaghan
Thames and Hudson, 1973

Microfilming for Beginners
By A D Smith
Business Equipment Trade Association, 1975

Motion Picture Camera and Lighting Equipment
By David Samuelson
Focal Press, 1977

Practical Motion Picture Photography
Edited by Russell Campbell
Tantivy Press

Script Continuity and the Production Secretary
By Avril Rowlands
Focal Press, 1977

Scriptwriting for Animation
By Stan Hayward
Focal Press, 1977

The Super 8 Film Makers' Handbook
By Myron A Matzin
Focal Press, 1976

The Technique of Film Animation
By John Halas and Roger Manvell
Focal Press, 1976

Visual Scripting
By John Halas
Focal Press, 1976

Work of the Industrial Film Maker
By John Burder
Focal Press, 1974

Work of the Science Film Maker
By Alex Strasser
Focal Press, 1972

Index of Advertisers

AV Distributors (London) Ltd, 13
Bell & Howell AV Ltd, 17
C W Cameron Ltd, 75
CEGB/Sudbury Conference Hall, 113
CPD – Consolidated Photovisual Distributors Ltd, 127
DRH Screens Ltd, 85
Electrosonic Ltd, 27
ESL Bristol, Training Services Div., 111
Film Distributors Associated Ltd, 99
Filmstrip Services Ltd, 116
Focal Point Audio Visual Ltd, 95
GDP Samuelson Sight & Sound (Birmingham) Ltd, 128
Gordon AV Ltd, 67
Francis Gregory & Son Ltd, 5
H & H Productions, 21
International Tutor Machines Ltd, 2, *outside back cover*
Jordan Hill College of Education, 120
Leeholme Audio Services Ltd, 107, 109
Livesey of Hull AV, 5
Lomax Audio Visual Division, 23
Media Facilities (Scotland) Ltd, 29
Midlands Video Systems Ltd, 73, 129
Multimedia Publishing Ltd, 97
Penrose Cine (Dollands) Ltd, 130
The Perforated Front Projection Company, 86
Reach-a-Teacha Ltd, 103
Robert Rigby Ltd, 89
Robian Video, 31, 115
Scottish Council for Educational Technology, 126
Tandberg (UK) Ltd, 90
Valiant Electrical Wholesale Company, 43, 109
Videogram Journal, 125
Visual Aids Centre, 19
D Wyllie Ltd, 69